Stolen from René T. Robichaux

A REFRESHER COURSE IN MATHEMATICS

F. J. Camm

Dover Publications, Inc.
Mineola, New York

Bibliographical Note

This Dover edition, first published in 2003, is an unabridged republication of the 1953 reprint by Emerson Books, New York, of the work originally published in London in 1943.

Library of Congress Cataloging-in-Publication Data

Camm, F. J. (Frederick James), 1897–1959.
A refresher course in mathematics / F.J. Camm.
p. cm.
Originally published: London : G. Newnes limited, 1943.
Includes index.
ISBN 0-486-43225-4 (pbk.)
1. Mathematics. I. Title.

QA113.C35 2003
510—dc21

2003055295

Manufactured in the United States of America
Dover Publications, Inc., 31 East 2nd Street, Mineola, N.Y. 11501

PREFACE

This book is intended to provide a refresher course in mathematics for those who have previously mastered the subject but have forgotten the fundamental facts concerning the various branches of calculation. It is not intended as a first course, and it has been presumed that the reader will recall by a perusal of this book his earlier instructions in the subject. None the less, it will be found easy to follow by those making their first acquaintance with mathematics.

I have received a large number of letters from readers of my journals asking me to prepare such a course so that they could study for the entrance examinations set by the various Services and Institutes associated with engineering and mechanics.

Although in one volume I have covered many of the branches which are normally comprised by several volumes, I have been enabled to do this by adopting the plan mentioned above, namely, of considering the book as a refresher course.

I have given sufficient examples to remind the reader of the various points associated with particular calculations, at least one example of each mathematical concept being given.

The book is arranged on the plan I found successful in teaching mathematics and machine drawing, and the chapters take the reader up to matriculation standard. Tables of trigonometrical functions and logarithms which the reader will find necessary in working some of the examples are included at the back of the book.

This fifth edition has been revised.

F. J. CAMM.

CONTENTS

CHAPTER I

MATHEMATICAL TERMS AND SIGNS

BEFORE we can use figures to arrive at correct results we must possess certain tools in the form of the standard *symbols*. These symbols are used all over the world and it is very necessary to memorise them, because they are, in effect, the shorthand of calculation. Here they are :

$+$ Plus, or add.
$-$ Minus, or subtract.
$\times$ Multiply by.
$\div$ Divide by.
$/$ Divide by (Solidus).
$=$ Is equal to.
$\equiv$ Is always equal to. Identical with.
$\backsimeq$, $\div$, or $\doteqdot$ Approximately equal to.
$\therefore$ Therefore.
$\because$ Since, because.
(Single bracket.
{ Double bracket, or brace.
[Square bracket.

$\sim$ Difference of.
$<$ Less than.
$>$ Greater than.
$\leqq$ Equal to, or less than.
$\geqq$ Equal to, or greater than.
$\nless$ Not less than.
$\ngtr$ Not greater than.
$\propto$ Varies as.
∞ Infinity.
$\parallel$ Parallel with.
$\perp$ Perpendicular to.
$\overline{}$ Vinculum or bar drawn over a group of algebraic terms (but the use of brackets is preferable).

$\pm$ Plus or minus, *i.e.* either plus or minus, according to circumstances.
$|\!+$ Modified plus sign, indicates that direction is taken into account as well as addition, as in obtaining the vector sum of two forces.
$\forall$ Sign of vector subtraction.

The symbols °, ′, and ″ are used to denote a degree, a minute, and a second respectively in the sexagesimal system of measurement, in which a right angle is divided into ninety equal parts called degrees, and each degree is divided into sixty equal parts called minutes, and each minute is divided into sixty equal parts called seconds.

Σ Sigma, the sum, or "summation of the products of."

π Pi, the ratio of circumference to diameter, also 180° in circular measure.

θ Theta, any angle from the horizontal.

ϕ Phi, any angle from the vertical.

⊙ Circle, or station point. △ Triangle, or trig station.

$\sqrt{\ }$ Square root. The $\sqrt{\ }$ sign is known as a radical. $\sqrt[3]{\ }$ Cube root. $\sqrt[4]{\ }$ Fourth root.

$\underline{|5}$ means factorial 5, or continued product up to $5=1\times2\times3\times4\times5$. The factorial sign is sometimes written thus : ! Hence 5 ! means the continued product up to 5, and is the same as $\underline{|5}$.

Π n=continued product of numbers up to $n=1\times2\times3\ \ldots\ n$.

a, b, c used for known quantities ; x, y, z for unknown quantities. n is used in place of any whole number.

A full stop (.) is sometimes used instead of the multiplication sign.

▭ Parallelogram.

□ Square.

○ Circumference.

δ Sign of differentiation.

$\neq$ Unequal to.

: Is to.

:: As ; so is (ratio)

∦ Not parallel.

∠ Angle ∟ Right angle.

◠ Semi-circle.

◔ Quadrant.

⌒ Arc.

( ), [ ], { } Vincula.

$\int_{o}^{x}$ is the sign of integration between limits o and x.

The symbol $\int_{a}^{b} y.dx$ means "the area beneath the curve whose ordinate is y, from x=a to x=b."

$\Sigma \sin^3 \theta \int_{0^\circ}^{90^\circ}$ means the summation of the cubes of the sines of the angles 0 to 90.

The sign + (plus) or — (minus) has a higher separating power in a formula than × (multiplication) ; therefore parts connected by × may be multiplied out before passing beyond the other signs. Remember that plus times plus =+, plus times minus= —, minus times minus=+.

ROMAN NUMERALS

1	2	3	4	5	6	7	8	9	10	50	100	500	1,000	1,000,000
I	II	III	IV	V	VI	VII	VIII	IX	X	L	C	D	M	$\overline{\text{M}}$

It will be observed that IV=4, means 1 short of 5 ; in the same way IX=9, means 1 short of 10 ; XL=40, means 10 short of 50 ; XC=90, means 10 short of 100 ; so for 1814 we have MDCCCXIV, and for 1953 we have MCMLIII.

OTHER MATHEMATICAL TERMS

Acnode.—A point outside a curve whose co-ordinate satisfies the equation of the curve ; a conjugate point.

Addend.—A quantity or number which is to be united in one sum with another quantity or number. This latter quantity is called the *augend.*

Adfected.—Containing different powers of an unknown quantity.

Aliquant.—Contained in another number but with a remainder.

Aliquot.—Contained in another number without a remainder.

Augend.—A quantity or number to which another is to be added. See *Addend.*

Binary Logarithms.—A system for use in musical calculations, in which 1 is the log. of 2, and the modulus is 1·442695.

Binary Scale.—The scale of notation whose ratio is 2, in which, therefore, 1 of the denary scale is 1, 2 is 10, 3 is 11, 4 is 100, etc.

Binary system.—The binary system of arithmetic is a method of computation in which the binary scale is used. Chiefly used in classification.

A *corollary* is a geometrical truth deducible from a theorem.

Cube (odd number).

Nought (0) is known as a *cypher.* *Digits* are the figures 1, 2, 3 4, 5, 6, 7, 8, 9, 0.

Denary System.—The decimal system in which 10 is the basis of calculation.

Dividend.—The number preceding the sign of division (÷) is called the *dividend.* The number following the sign of division is called the *divisor.* The answer to a division sum is called the *quotient.*

Ellipse.—A plane closed curve in which the sum of the distances of any point from the foci is a constant quantity.

An *equilateral triangle* has three equal sides.

Exponent.—Same as *index.*

A *factor, multiple,* or *measure,* is a number which divides an exact number of times into any other number. Sometimes it is referred to as a *sub-multiple* of the number.

Greatest Common Factor (see *Highest Common Factor*).

The *Highest Common Factor* (H.C.F.), sometimes termed the *Greatest Common Measure*—or *Greatest Common Divisor,* is the greatest number which is contained an exact number of times in each of two or more numbers. It is usually abbreviated to H.C.F. The usual method of finding the H.C.F. of two or more numbers is to break each number up into its factors.

Taking the two numbers 15 and 75, the factors are :

15 equals 3×5.
75 equals $3\times5\times5$.

The H.C.F. is therefore $3\times5=15$.

Another example : find the H.C.F. of 5005 and 2805.
The factors of 5005 are $5\times7\times11\times13$.
The factors of 2805 are $3\times5\times11\times17$.

The H.C.F. is 5×11, which equals 55.

If a number contains another number an exact number of times, the first number is termed a *multiple* of the second. The smallest number which contains each of two or more numbers is known as the *Least Common Multiple,* abbreviated to L.C.M. It will be observed that in factorising, the factors are *prime* numbers. To find the L.C.M. of two or more numbers, each of the numbers is split up into its *prime factors.* When this is done, choose the highest powers of each prime factor which occur in the products. The L.C.M. is obtained by multiplying the highest powers together. Thus, the L.C.M. of 110, 45, and 40 is found in the following way :

110 equals $2 \times 5 \times 11$.
45 equals $3^2 \times 5$.
40 equals $2^3 \times 5$.

The highest powers of 2, 3, 5, and 11 which occur are 2^3, 3^2, 5, and 11. Therefore, the L.C.M. is $2^3 \times 3^2 \times 5 \times 11$, which equals $8 \times 9 \times 5 \times 11$, or 3960.

Index.—The number denoting the power of a given quantity.

Any whole number is known as an *integer*.

Irrational or *indeterminate* (see *Surd*).

An *isosceles triangle* has two equal sides.

Measure (see *Factor*).

Minuend.—The number from which another is to be subtracted, opposed to *subtrahend* (which see).

Multiple (see *Factor*).

Multiplicand.—The number preceding the sign of multiplication ($\times$) is known as the *multiplicand*. The number following the sign of multiplication is called the *multiplier*. The result of multiplication is called the *product*. The two numbers themselves (the number before the multiplication sign and the number after it) are called *factors of the product*.

Operand.—Any quantity or symbol to be operated on. A *Faciend*.

A *parallelogram* is a four-sided figure having opposite sides *parallel*.

A *postulate* is something to be done of which the possibility is admitted.

A *prime number* is any number which is not divisible by any other number except *unity* (1). Hence 1, 3, 5, 7, 11, 13, 17, 19, etc., are prime numbers.

Product (see *Multiplicand*).

Q.E.D. stands for *quod erat demonstrandum*. This is placed at the end of a *theorem* to mark that the truth of it has been proved (popularly, *quite easily done !*).

Q.E.F. is placed at the end of a *problem*, and stands for *quod erat faciendum*, to indicate that the problem has been done.

A *quadrilateral* is any figure enclosed by four straight lines.

Quartic.—A rational homogeneous function of any number of variables.

Quaternion.—The quotient of two vectors, or the operator which changes one vector into another, so called because it depends on four geometrical elements and is capable of being expressed by the quadrinomial formula:

$$w+xi+yj+2zk,$$

in which w, x, y, z are scalars, and i, j and k are mutually perpendicular vectors whose squares are -1.

Radix.—A quantity regarded as a base or fundamental unit, as 10 is the radix of the common system of logarithms.

Solidus.—The division sign (/).

A *reciprocal* is the quotient obtained by dividing unity by a number. Thus the reciprocal of 2 is $\frac{1}{2}$, and the reciprocal of $\frac{1}{2}$ is 2.

A *rectangle* is a parallelogram having each of its angles right angles, and its opposite sides equal in length. Thus, an *oblong* is a rectangle.

Recurring decimal (see *Surd*).

A *reflex angle* is one which is greater than two right angles.

Repetend.—That part of a circulating or recurring decimal which is repeated indefinitely.

A *rhombus*, or *rhomb*, is a parallelogram with all its sides equal but none of its angles is a right angle.

A *right-angle triangle* has one angle of 90°, and the side opposite the right angle is called the *hypotenuse*.

Scalar.—A pure or real number; that term of a quaternion which is not a vector, but a real number.

A *scalene triangle* has three unequal sides.

A *square* is a parallelogram which has all its sides equal and all its angles right angles.

Square.—When any number is multiplied by itself the result is known as the *square*, or second power of the number. When the number is multiplied by itself twice, the result is the *cube* or third power of the number. If any number is multiplied by itself three times, the result is the *fourth power*, and so on.

A method of indicating the power of a number is to put a small figure near the right-hand top of the figure; thus, 7^3 means 7 raised to the third power, or 343, which equals $7\times7\times7$.

Subtrahend.—That which is to be subtracted, opposed to *minuend.*

A *surd* is a quantity which is *irrational* or *indeterminate.* Thus a *recurring decimal* is a surd, and so is any unending decimal, like *pi.* Such numbers are sometimes called *incommensurable numbers.*

Tangent.—(Geometry) A line touching, but not intersecting, a curved line or surface; (Trigonometry), one of the three fundamental functions of an angle (*see* Chapter XXIII).

A *theorem* is a truth capable of demonstration by reasoning of known truths.

A *trapezium* is a four-sided figure which has only two of its sides parallel.

A *triangle* is a plane figure bounded by three straight lines. Any one of its points may be considered as the *vertex* ; the opposite side to the vertex is called the *base*, and the altitude of the triangle is the perpendicular distance of the vertex from the base.

A *vector* is a line whose length represents the amount of a quantity, and whose direction indicates which way the quantity is acting.

Vertex (see *Triangle*).

CHAPTER II

FRACTIONS

A *fraction* consists of two numbers separated by a horizontal line. The one above the line is known as the *numerator*, and the one below is the *denominator*. Thus, $\frac{1}{2}$, $\frac{1}{4}$, $\frac{1}{8}$, $\frac{3}{4}$, $\frac{3}{8}$, $\frac{5}{8}$, $\frac{7}{8}$ are fractions. The denominator indicates the number of parts the *unit* has been divided into, whilst the numerator indicates the number of those parts we are considering. Thus half a crown is $\frac{1}{8}$th of a £, but in relation to 5s. it is $\frac{1}{2}$. There are various sorts of fractions. A *proper* fraction is one which is less than unity (1). The fractions given above are all proper fractions. An *improper* fraction is one which is more than unity, such as $\frac{27}{6}$, $\frac{13}{4}$.

When the numerator and denominator of a fraction are each multiplied or divided by the same number the value of the fraction is unaltered. For example, $\frac{1}{4}$ is the same as $\frac{4}{16}$, or $\frac{25}{100}$.

Addition.—When adding, subtracting, or comparing fractions each having the same denominator, it is only necessary to add, subtract, or compare the numerator of such fractions, but when the denominators are not similar the fractions must first be reduced to equivalent fractions each having the same denominator. Hence, to add $\frac{3}{7}$ths and $\frac{2}{7}$ths together it is merely necessary to add the numerators $=\frac{5}{7}$ths. In order to add together $\frac{3}{4}$ and $\frac{1}{16}$ we must convert $\frac{3}{4}$ into 16ths, by multiplying the numerator and denominator by 4, producing $\frac{12}{16}$ths, which added to the $\frac{1}{16}$th gives $\frac{13}{16}$ths as the answer. Or again, to add together $\frac{3}{7}$ and $\frac{5}{9}$ we can multiply the denominators together, making 63rds. Hence, $\frac{3}{7}=\frac{27}{63}$, and $\frac{5}{9}=\frac{35}{63}$. We can now add the numerators together, making $\frac{62}{63}$. We could obtain exactly the same result by what is known as cross multiplication, thus taking the two fractions $\frac{3}{7}$ and $\frac{5}{9}$ we obtain 9 times $3=27$, and 7 times $5=35$. We now multiply the denominators together, thus obtaining $\frac{27}{63}+\frac{35}{63}$, which $=\frac{62}{63}$. It is always necessary when adding fractions to convert each of them to a *common denominator*. It is important to note that any whole number can always be expressed in a fractional form, thus $9=\frac{9}{1}$.

A *mixed number* is one consisting of a whole number and a fraction, thus $5\frac{1}{2}$, $6\frac{1}{3}$, $2\frac{1}{8}$, are mixed numbers. These can be expressed as fractions : $5\frac{1}{2}=\frac{11}{2}$, $2\frac{1}{8}=\frac{17}{8}$, and $6\frac{1}{3}=\frac{19}{3}$. Observe that in order to convert a mixed number into a fraction the denominator is multiplied by the whole number and the numerator added.

It is often convenient when adding up mixed numbers to add the whole numbers first, but in multiplication and division it is best to reduce the mixed number to an improper fraction.

Multiplying and Dividing Fractions.—To multiply a fraction by a number, either *multiply its numerator by the number or divide its denominator by the number.* For example, multiply $\frac{5}{16}$ by 4. Multiplying the numerator we get $\frac{20}{16}$, which equals $1\frac{1}{4}$. Alternatively, divide 16 by 4 and so obtain $\frac{5}{4}$. To divide a fraction by a whole number, either *divide the numerator by the number or multiply the denominator by it.* Example : Divide $\frac{3}{16}$ by 4. Multiplying the denominator by 4 we get $\frac{3}{64}$. But $\frac{3}{16}=\frac{12}{64}$, and dividing 12 by 4 we obtain $\frac{3}{64}$ in the same way.

To multiply a fraction by a fraction, multiply the numerators together and the denominators together. Thus, $\frac{5}{8}\times\frac{3}{4}=\frac{15}{32}$.

Cancelling.—When a number of fractions have to be multiplied together *the numerators are multiplied together and also the product of the denominators providing the numerator and the denominator of the result.* Before doing this, it is useful to cancel out as far as possible in order to avoid unwieldy multiplication sums and final cancelling. For example :

$$\frac{3}{4}\times\frac{\overset{4}{\cancel{32}}}{\cancel{8}}\times\frac{\cancel{23}}{\cancel{16}}\times\frac{\cancel{16}}{\cancel{23}}\times\frac{\cancel{5}}{\cancel{4}}\times\frac{3}{\underset{4}{\cancel{20}}}=\frac{3}{4}\times\frac{3}{4}=\frac{9}{16}$$

From this it will be noticed that the 23's and the 16's cancel out, numerator 5 divides into the 20, leaving 4, denominator 8 divides into numerator 32, leaving 4, which cancels out the denominator of the fraction $\frac{5}{4}$, leaving $\frac{3}{4}\times\frac{3}{4}$, which equals $\frac{9}{16}$.

Dividing One Fraction by Another.—To divide one fraction by another invert the divisor and multiply the remaining fraction by it. Example : Divide $\frac{9}{32}$ by $\frac{27}{16}$. Inverting $\frac{27}{16}$ we get $\frac{16}{27}$, and multiplying $\frac{9}{32}$ by $\frac{16}{27}$ obtain $\frac{144}{864}$. This will cancel down to $\frac{1}{6}$, because 864 is exactly six times 144. We could, and in practice would, have arrived at this result by cancelling the fractions themselves, thus :

$$\frac{\not{9}}{\not{32}_{2}}\times\frac{\not{16}}{\not{27}_{3}}=\frac{1}{2}\times\frac{1}{3}=\frac{1}{6}$$

Further Examples.—Divide $3\frac{1}{8}$ by $2\frac{1}{4}$.

$$\frac{25}{\not{8}_{2}}\times\frac{\not{4}}{9}=\frac{25}{18}$$

Divide $\frac{1}{10}$ by $7\frac{1}{2}$. Answer, $\frac{1}{75}$.
Divide $1\frac{1}{17}$ by $\frac{2}{51}$. Answer, 27.
Multiply $1\frac{1}{17}$ by $\frac{2}{15}$. Answer, $\frac{12}{85}$.
Multiply $\frac{1}{16}$ by $7\frac{1}{2}$. Answer, $\frac{15}{32}$.
Multiply $4\frac{3}{16}$ by $8\frac{1}{3}$. Answer, $34\frac{43}{48}$.
Divide $4\frac{3}{16}$ by $8\frac{1}{3}$. Answer, $\frac{201}{400}$.
Divide $\frac{1}{96}$ by $3\frac{1}{8}$. Answer, $\frac{1}{300}$.
Multiply $\frac{7}{8}$ by $\frac{3}{16}$. Answer, $\frac{21}{128}$.

CHAPTER III

Continued Fractions: Approximations

Quite often in workshop calculations an unwieldy fraction occurs. A fraction having a large denominator consisting of a prime number is reduced to simpler terms by means of continued fractions. In this system of continued fractions unity is used for the numerator and the denominator is an entire number or integer plus a fraction.

Continued fractions, therefore, enable us to find another fraction expressed in smaller numbers which may be sufficiently approximate to another value expressed in large numbers.

Example: What fraction expressed in smaller numbers is nearest in value to $\frac{29}{146}$? Dividing the numerator and denominator by the same number does not change the value of the fraction. Dividing both terms of $\frac{29}{146}$ by 29, we have $\dfrac{1}{5\frac{1}{29}}$, or, what is the same thing expressed as a continued fraction, $\dfrac{1}{5+\frac{1}{29}}$. The continued fraction $\dfrac{1}{5+\frac{1}{29}}$ is exactly equal to $\frac{29}{146}$. If, now, we reject the $\frac{1}{29}$, the fraction $\frac{1}{5}$ will be larger than $\dfrac{1}{5+\frac{1}{29}}$, because the denominator has been diminished, 5 being less than $5\frac{1}{29}$. $\frac{1}{5}$ is something near $\frac{29}{146}$ expressed in smaller numbers than 29 for a numerator and 146 for a denominator. Reducing $\frac{1}{5}$ and $\frac{29}{146}$ to a common denominator, we have $\frac{1}{5}=\frac{146}{730}$ and $\frac{29}{146}=\frac{145}{730}$. Subtracting one from the other we have $\frac{1}{730}$, which is the difference between $\frac{1}{5}$ and $\frac{29}{146}$.

There are fourteen fractions with terms smaller than 29 and 146, which are nearer $\frac{29}{146}$ than $\frac{1}{5}$ is, such as $\frac{15}{76}$, $\frac{16}{81}$, and so on to $\frac{28}{141}$. In this case by continued fractions we obtain one approximation, namely $\frac{1}{5}$, and any other approximations, such as $\frac{15}{76}$, $\frac{16}{81}$, etc., we find by trial. It will be noted that all these approximations are smaller in value than $\frac{29}{146}$. There are cases, however, in which we can, by continued fractions, obtain approximations both greater and less than the required fraction.

In the French metric system, a millimetre is equal to ·03937 inch : What fraction in smaller terms nearly expresses ·03937 inch ? ·03937, in a vulgar fraction, is $\frac{3937}{100000}$. Dividing both numerator and denominator by 3937, we have $\frac{1}{25+\frac{1575}{3937}}$. Rejecting from the denominator of the new fraction, $\frac{1575}{3937}$, the fraction $\frac{1}{25}$ gives us a pretty good idea of the value of ·03937 inch. If, in the expression $\frac{1}{25\frac{1575}{3937}}$, we divide both terms of the fraction $\frac{1575}{3937}$ by 1575, the value will not be changed. Performing the division we have

$$\cfrac{1}{25+\cfrac{1}{2+\cfrac{787}{1575}}}$$

We can now divide both terms of $\frac{787}{1575}$ by 787, without changing its value, and then substitute the new fraction for $\frac{787}{1575}$ in the continued fraction.

Dividing again, and substituting, we have :

$$\cfrac{1}{25+\cfrac{1}{2+\cfrac{1}{2+\cfrac{1}{787}}}}$$

as the continued fraction that is exactly equal to ·03937.

In performing the divisions, the work stands thus :

```
3937)100000(25
      7874
      21260
      19685
       1575)3937(2
            3150
             787)1575(2
                 1574
                    1)787(787
                      787
                        0
```

That is, dividing the last divisor by the last remainder, as in finding the greatest common divisor, the quotients become the denominators of the continued fraction, with unity for numerators. The denominators 25, 2, and so on, are called incomplete quotients, since they are only the entire parts of each quotient. The first expression in the continued fraction is $\frac{1}{25}$ or ·04—a little larger than ·03937. If, now, we take $\frac{1}{25+\frac{1}{2}}$, we shall come still nearer ·03937. The expression $\frac{1}{25+\frac{1}{2}}$ is merely stating that 1 is to be divided by $25\frac{1}{2}$. To divide, we first reduce $25\frac{1}{2}$ to an improper fraction, $\frac{51}{2}$, and the expression becomes $\frac{1}{\frac{51}{2}}$, or one divided by $\frac{51}{2}$. To divide by a fraction, invert the divisor, and proceed as in multiplication. We then have $\frac{2}{51}$ as the next nearest fraction ·03937. $\frac{2}{51}$=·0392+, which is smaller than ·03937. To get still nearer, we take in the next part of the continued fraction, and have :

$$\cfrac{1}{25+\cfrac{1}{2+\cfrac{1}{2}}}$$

We can bring the value of this expression into a fraction, with only one number for its numerator and one number for its denominator, by performing the operations indicated, step by step, commencing at the last part of the continued fraction. Thus, $2+\frac{1}{2}$ or $2\frac{1}{2}$ is equal to $\frac{5}{2}$. Stopping here, the continued fraction would become—

$$\cfrac{1}{25+\cfrac{1}{\cfrac{5}{2}}}$$

Now, $\frac{1}{\frac{5}{2}}$ equals $\frac{2}{5}$, and we have $\frac{1}{25+\frac{2}{5}}$. $25\frac{2}{5}$ equals $\frac{127}{5}$; substituting again, we have $\frac{1}{\frac{127}{5}}$. Dividing 1 by $\frac{127}{5}$, we have $\frac{5}{127}$. This is the nearest fraction to ·03937, unless we reduce the whole continued fraction :

$$\cfrac{1}{25+\cfrac{1}{2+\cfrac{1}{2+\cfrac{1}{787}}}}$$

which would give us back the ·03937 itself.

$\frac{5}{127}=\cdot 03937007$, which is only $\frac{7}{100000000}$ larger than ·03937. It is not often that an approximation will come so near as this.

This ratio, 5 to 127, is used cutting millimetre thread screws. If the leading screw of the lathe is 1 to 1 in., the change gears will have the ratio of 5 to 127 ; if 8 to 1 in., the ratio will be 8 times as large, or 40 to 127 ; so that with leading screw 8 to inch, and change gears 40 and 127, we can cut millimetre threads sufficiently accurate for practical purposes.

CHAPTER IV

Decimals

Definitions.—Any fraction in which the denominator is 10, or some multiple of 10, is known as a *decimal fraction.* We know that in whole numbers each digit to the left increases in value by ten times. Thus the number 6,843,291 means

1 unit
9 tens
2 hundreds
3 thousands
4 ten thousands
8 one-hundred thousands
6 millions

Each move to the left signifies a value ten times greater than that of the place preceding it. Conversely, each move to the right reduces the value of the digits to one-tenth.

This system is the basis of *decimals*, and it is also employed to express numbers which are less than unity. We use a *decimal point* to separate the whole numbers from the fraction, putting the whole numbers (if any) to the left of the decimal point, and the fractional number to the right of it. Thus 192·375 means one hundred and ninety-two whole units, plus $\frac{3}{10}$ths of a unit, plus $\frac{7}{100}$ths of a unit, plus $\frac{5}{1000}$ths of a unit, and such a decimal would be expressed verbally as *one hundred and ninety-two, point three seven five.*

Subtraction and Addition of Decimals.—The ordinary rules of arithmetical multiplication and division apply to decimals, but it is important to *keep the decimal points of the quantities being added or subtracted under one another.* Here is an example of decimal addition:

$$\begin{array}{r} 39{\cdot}0625 \\ 14{\cdot}31975 \\ 2{\cdot}47113 \\ 125{\cdot}00139 \\ \hline 180{\cdot}85477 \\ \hline \end{array}$$

Subtraction is, of course, the reverse :

```
 187·923875
  63·198362
 ----------
 124·725513
 ----------
```

Multiplication of Decimals.—In multiplication *treat the two quantities as whole numbers.* The position of the decimal point is obtained *by counting the number of fractional digits in the multiplicand and the multiplier, adding these together, and counting off from the right of the product this number of digits ;* the decimal point is placed to the left of the digit so counted off. For example : Multiply together 39·675 by 84·2163.

```
    84.2163
    39·675
 -----------
    4210815
   58951410
  505297800
 7579467000
25264890000
 -----------
33412817025
 -----------
```

There are four decimal places in one quantity, and three in the other : $4+3=7$, so there will be seven decimal places in the product, and the latter thus becomes, counting off seven places *from the right,*

3341·2817025

The same method applies irrespective of the number of decimal quantities which are to be multiplied together.

Decimal Approximations. — Now, 3341·2817025 for all practical purposes can be shortened to 3341·282, rejecting the remainder of the decimal. Such a shortened result is said to be approximately *correct to three decimal places or to three significant figures.* There is a rule concerning this shortening process. If a rejected or discarded decimal is 5 or over, one is added to the next figure to the left. Thus, in the decimal given above, 7 being greater than 5 is rejected, and the figure 1 to the left of it is increased to 2. Another method of decimal approximation which has been approved internationally is to *make the decimals even.* That is to say, in the case of (for example) the number 39·455, we should shorten this to 39·46 (adding 1 for the 5 which is dropped or for any number over 5). In the case of a number such as 39·444 we should merely shorten this to 39·44. Making the decimals even, it is claimed, gives a closer average result.

If the result were required approximately correct to one decimal place, we should shorten 3341·2817025 to 3341·3. If required correct to one, two, three, etc., decimal places, the decimals beyond are merely discarded, thus :

3341·281
3341·28
3341·2

are correct to three, two, and one decimal places respectively.

If the decimal is purely fractional and contains a number of noughts after the decimal point, at least one significant figure must, of course, be left. For example, in the decimal fraction—

·0000063192

we may shorten it only to ·000006.

When multiplying decimals by 10, or any multiple or submultiple of 10, *it is merely necessary to move the decimal point* one place for each power of 10 in the multiplier.

Example : Multiply 394·264 by 100.
Answer : 39426·4 (the decimal point moved two places because $100 = 10 \times 10$).

In multiplying by any multiple or submultiple of 10, the decimal point is moved, as in the above example, *to the right*, if the decimal includes a whole number. If the decimal is purely fractional, the decimal point is also moved *to the right*.

Example : Multiply ·38624 by 100.
Answer : 38·624.

The number of places the decimal point must be moved when multiplying by functions of 10 is *to count the number of digits in the multiplier and subtract* 1 *from it*. Thus :

To multiply by 10, move decimal point one place (2—1) ;

To multiply by 100, move the decimal point two places (3—1) ;

To multiply by 1000, move decimal point three places (4—1) ;

To multiply by 10,000, move decimal point four places (5—1) ;

and so on.

Division of Decimals.—In dividing decimals by 10 or multiples of 10, move the decimal point *to the left*—reversing the process explained above.

Example : Divide ·375 by 100.
Answer : .00375.
Divide 375·625 by 100.
Answer : 3·75625.

Division of decimals is carried out in the same manner as for whole numbers, and the fixing of the decimal point is the only part of the process which needs explanation. For example : divide ·95 by ·235. In other words, we must find a number which, when multiplied by ·235, produces ·95.

It is nearly always convenient to multiply the number in the divisor by some multiple of 10 which will make it a whole number. Thus, in the example given, $\cdot 235 \times 1000$ produces 235. We must, of course, multiply the dividend also by 1000, thus producing 950. Division is then carried out in the following way :

```
235)950·0(4·04255
    940
    ---
     10·00
      9 40
      ----
       60·0
       47 0
       ----
       130·0
       117 5
       -----
         125·0
         117 5
         -----  . . .
```

The same method is adopted if the divisor includes whole numbers as well as a decimal fraction. For example, divide 75·00625 by 3·125. Here we multiply the divisor by 1000, producing 3125 ; multiplying the dividend also by 1000 we obtain 75006·25. The division is then carried out in the following way :

```
3125)75006·25(24·002
     6250
     ─────
     12506
     12500
     ─────
         6250
         6250
         ────
         0000
```

Recurring Decimals.—In some cases of decimal division the calculation can be carried on indefinitely, and such decimals are known as *recurring* decimals.

For practical purposes it is not necessary to carry calculations beyond three places of decimals, and usually two or three significant figures suffice. It is important to remember that noughts immediately after the decimal point do not count as significant figures. Thus, in decimals such as ·0000329, the *first significant figure* is 3, and expressed correct to one significant figure the fraction is expressed as ·00003 ; to two significant figures ·000032, and so on.

It is always wise to discard the unnecessary figures, because they make the calculation unnecessarily lengthy and add to the possibility of error. Remember that for approximate results any figure over 5 may be added as 1 to the next decimal place *to the left*. The above decimal could thus be written (approximately correct) as ·000033.

Recurring, circulating, or repeating decimals are denoted by a dot over the recurring figure ; thus, ·30$\dot{3}$ means ·3033333 . . . and so on to infinity. Similarly, groups of figures in the decimal fraction may recur. Thus, ·$\dot{3}\dot{9}$3939, or ·7353$\dot{7}$35$\dot{3}$. In this case the dots are placed over the first and last figures of the recurring group. I shall deal with recurring decimals later, in connection with the conversion of fractions into decimals.

Conversion of Decimals to Vulgar Fractions.—It has already been explained that decimals represent tenths, hundredths, thousandths, etc., according to the position of the figure from the decimal point.

Thus $\cdot 5=\frac{5}{10}$, $\cdot 75=\frac{75}{100}$, $\cdot 375=\frac{375}{1000}$, $\cdot 0375=\frac{375}{10,000}$, and so on.

In order to convert a decimal into a fraction, *it is only necessary to use it as a numerator, with a denominator of* 1 *followed by as many noughts as there are decimal places in the fraction.* The above examples make this clear. Cancelling can then take place in the usual way. Examples :

$\frac{375}{10,000}=\frac{3}{80}$ (both numerator and denominator divided by 125).

$\frac{75}{100}=\frac{3}{4}$ (both divided by 25).

$\frac{5}{10}=\frac{1}{2}$ (both divided by 5).

Some fractions and their decimal equivalents occur so often in calculations that they should be committed to memory. I give them on page 27.

It should be noted that once the decimal equivalents of $\frac{1}{2}$, $\frac{1}{4}$, $\frac{1}{8}$, $\frac{1}{16}$, $\frac{1}{32}$, and $\frac{1}{64}$ have been memorised, it is an easy matter to find other fractional equivalents in this series. Thus for $\frac{3}{64}$ merely multiply ·015625 by 3. Similarly as $\frac{1}{4}=\cdot 25$, $\frac{1}{8}$=half of that decimal=·125.

Converting Fractions to Decimals.—Reversing the process, vulgar fractions may be converted into decimals by reducing them to their lowest terms, *and then dividing the numerator by the denominator.* Examples :

$\frac{5}{16}$=16)5·0(·3125

```
    4 8
    ---
     20
     16
     --
      40
      32
      --
       80
       80
       ==
```

$\frac{19}{37}$=37)19·0($\cdot 5\dot{1}3\dot{5}$

```
    18 5
    ----
      50
      37
     ---
     130
     111
     ---
      190
      185
      ---
        5
      ===
```

Table of Decimal Equivalents

Fraction	Decimal	Fraction	Decimal	Fraction	Decimal
1/64	·015625	25/64	·390625	45/64	·703125
1/32	·03125	13/32	·40625	23/32	·71875
3/64	·046875	27/64	·421875	47/64	·734375
1/16	·0625	7/16	·4375	3/4	·7500
5/64	·078125	29/64	·453125	49/64	·765625
3/32	·09375	15/32	·46875	25/32	·78125
7/64	·109375	31/64	·484375	51/64	·796875
1/8	·1250	1/2	·5000	13/16	·8125
9/64	·140625				
5/32	·15625				
11/64	·171875				
3/16	·1875				
13/64	·203125	33/64	·515625	53/64	·828125
7/32	·21875	17/32	·53125	27/32	·84375
15/64	·234375	35/64	·546875	55/64	·859375
1/4	·2500	9/16	·5625	7/8	·8750
17/64	·265625	37/64	·578125	57/64	·89062
9/32	·28125	19/32	·59375	29/32	·90625
19/64	·296875	39/64	·609375	59/64	·921875
5/16	·3125	5/8	·6250	15/16	·9375
21/64	·328125	41/64	·640625	61/64	·953125
11/32	·34375	21/32	·65625	31/32	·96875
23/64	·359375	43/64	·671875	63/64	·984375
3/8	·375	11/16	·6875	1	1·0000

Sometimes, by multiplying the numerator and denominator by a suitable number, it is possible to produce the decimal equivalent without division. Thus:

$$\frac{4}{25}=\frac{16}{100}=\cdot 16$$

$$\frac{23}{125}=\frac{184}{1000}=\cdot 184$$

Converting Recurring Decimals to Fractions.—Instead of using noughts as a denominator, *nines are used when converting recurring decimals into fractions.* Thus:

$$\cdot\dot{3}=\frac{3}{9}=\frac{1}{3}$$

$$\tfrac{1}{7}=\cdot\dot{1}4285\dot{7}\dot{1}4285\dot{7}\dot{1}4285\dot{7}\ \ldots\ldots$$

$=\frac{142857}{999999}$ (one 9 for every decimal place); 142857 divides or cancels into 999999 seven times$=\frac{1}{7}$.

These are *pure recurring decimals.* In the case of mixed recurring decimals, in which the decimal point is followed by some figures which do not recur, the rule is: *Subtract the non-recurring figures from all the figures, using the answer as the numerator, and for the denominator use as many nines as there are recurring figures, followed by as many noughts as there are non-recurring figures.* Examples:

$$\cdot 01\dot{9}\dot{6}=\frac{196-1}{9900}=\frac{195}{9900}=\frac{13}{660}$$

CHAPTER V

Duodecimals

In calculations involving feet and inches a system of computation known as duodecimals is used. As calculations in the building trades are almost entirely confined to feet and inches, duodecimals are very generally used by surveyors, architects, bricklayers, painters, and glaziers.

In the normal system of calculation, in order to multiply feet by inches it is necessary to reduce the feet to inches, and, similarly, if it is desired to multiply, say, 3 ft. 6 in. by 7 ft. 9 in., we should proceed to reduce each to inches, multiplying the two quantities together in the ordinary way, and then divide the answer by 144 to obtain the answer in square feet (144 sq. in. equals 1 sq. ft., and 1728 cu. in. equals 1 cu. ft.).

This somewhat cumbrous method is eliminated by the use of duodecimals. In this system the unit is the foot, and this is subdivided into 12 *primes* (1^{i}), the primes are subdivided in 12 *seconds* (1^{ii}), and the seconds are divided in 12 *thirds* (1^{iii}). In some textbooks the term *parts* is used in place of seconds.

Thus, the divisions and subdivisions of the square foot are termed *superficial primes*, *superficial seconds*, *superficial thirds*, etc. Similarly, in cubic measure the subdivisions are termed *cubic primes*, *cubic seconds*, *cubic thirds*, etc.

It will thus be seen that when primes are multiplied by feet the answer is in *superficial primes*, the product of feet times feet being, of course, square feet. Hence, one prime equals $^1/_{12}$ sq. ft., and therefore one prime multiplied by 1 ft. equals $^1/_{12} \times 1$, which equals one *superficial prime*. From this it will be seen that by multiplying feet and seconds the product will be termed *superficial seconds*.

Here is an example :

Multiply 7 ft. 6 in. by 5 ft. 8 in.; then multiply the product by 8ft. $2\frac{1}{2}$ in.

First write down the two quantities to be multiplied

together, putting like quantities immediately beneath each other.

Commence with the number indicating feet in the multiple, and use the lower dimensions as the multiplicand, carrying 1 to the left for each 12 in each product.

Now multiply by the inches, carrying forward the result of the mental division by 12 and setting the result down to the right of the former product. Proceed in this manner if there is a third dimension. Thus :

5 ft.	8^{I}	
7	6^{I}	
39	8^{I}	
2	10^{I}	0^{II}
42 ft.	6^{I}	0^{II}

In this example $7 \times 8 = 56$. Dividing 56 mentally by 12 we obtain $4 + 8$. Set down the 8 and carry the 4. Now $7 \times 5 = 35$, and $35 + 4 = 39$.

Next we multiply by 6, obtaining $6 \times 8 = 48$, and this divided by $12 = 4$. Hence we put 0 down and carry forward 4. So $6 \times 5 = 30$ and adding the $4 = 34$, and 12 divided into this provides 2 ft. 10^{I}. Adding these together, we obtain as a result 42 ft. 6^{I} 0^{II}.

In multiplying by the third dimension (8 ft. $2\frac{1}{2}$ in.) this dimension, in duodecimals, is 8 ft. $+ 2^{6}/_{12}$ in. $=$ 8 ft. 2^{I} 6^{II}.

Multiply as before :

42 ft.	6^{I}	0^{II}		
8	2	6		
340	0^{I}	0^{II}		
7	1^{I}	0^{II}	0^{III}	
1	9^{I}	3^{II}	0^{III}	0^{IV}
348 ft.	10^{I}	3^{II}	0^{III}	0^{IV}

$$i.e.\ 348 \text{ cu. ft.} + \left(\frac{10}{12} + \frac{3}{144}\right) \times 1728$$

$$= 348 \text{ cu. ft.} + \left(\frac{120}{144} + \frac{3}{144}\right) \times 1728$$

$$= \mathbf{348 \text{ cu. ft.} + 1476 \text{ cu. in.}}$$

Another example : Multiply 5 ft. 8 in. $\times$ 3 ft. 4 in.

3 ft.	4^{i}	
5	8	
16 ft.	8^{i}	
2	2^{i}	8^{ii}
18 ft.	10^{i}	8^{ii}

Which equals 18 sq. ft. $+\left(\frac{10}{12}+\frac{8}{144}\right)\times 144$

$=18$ sq. ft. $+\left(\frac{120}{144}+\frac{8}{144}\right)\times 144$

$=18$ sq. ft. $+128$ sq. in.

Division is carried out as in this example :

Divide 16 sq. ft. 90 sq. in. by 3 ft. 6 in.

3 ft. 6 in.)	16 sq. ft. 7^{i} 6^{ii}	(4 sq. ft. 9 sq. in.
	14 sq. ft. 0^{i}	
	2 $\quad 7^{i}$ 6^{ii}	
	2 $\quad 7^{i}$ 6^{ii}	

It will be seen that 4 is tried as a multiplier : $4\times 6=24$. Put down 0 and carry 2 ; $4\times 3=12$; add 2 and obtain 14. Subtract, thus getting 2.7^{i} ; bring down 6^{ii}. Try 9 : $9\times 6=54$; put down 6^{ii} and carry 4 ; $9\times 3=27$; add 4 and obtain 31 ; $31=2.7^{i}$.

CHAPTER VI

Square Root and Cube Root

Square Root.—The method of extracting square root is as follows :

Mark off the number, the square root of which is to be found, into periods by marking a dot over every second figure starting at the decimal point. Draw a vertical line to the left of the figure and a bracket on the right-hand side. Next, find the largest square in the left-hand period, and place this root behind the bracket. Next, the square of this root is subtracted from the first period, and the next period is brought down adjacent to the remainder and used as a dividend. Now multiply the first root found by 2 and place this product to the left of the vertical line ; then divide it into the left-hand figures of this new dividend, ignoring the right-hand figure. Attach the figure thus obtained to the root, and also to the divisor. Multiply this latest divisor by the figure of the root last obtained, finally subtracting the product from the dividend. Continue this operation until all the periods have been brought down. If a decimal fraction is involved, the periods for the decimal are marked off to the right of the decimal point.

The following examples will make the process clear. The first trial divisors are underlined in each case.

Example : Find the square root of 1156.

```
    ·  ·
 3|1156(34
  | 9
  | —
64| 256
- | 256
  |———
```

Find the square root of 54756 :

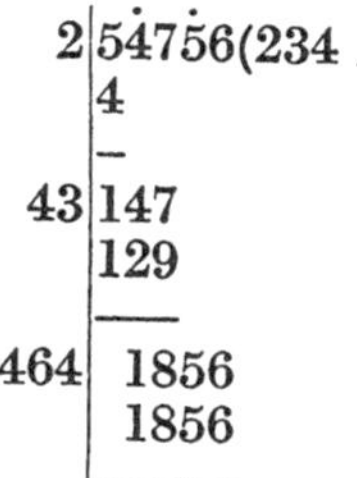

```
    2 | 54756(234
      | 4
      | -
   43 | 147
      | 129
      | ---
  464 |  1856
      |  1856
      | ----
```

Find the square root of 39·476089 :

```
     6 | 39·476089(6·283
       | 36
       | --
   122 |  347
       |  244
       |  ---
  1248 |  10360
       |   9984
       |  -----
 12563 |    37689
       |    37689
       |    -----
       |  . . . . .
```

Cube Root.—Separate the given number into periods by placing a dot over every third figure above the decimal point.

Determine the greatest number whose cube does not exceed the number in the left-hand period. This number is the first part of the required cube root. It is here called a "partial root." Subtract its cube from the left-hand period and to the remainder bring down the next period to form a second dividend. To form the next divisor, first form a partial divisor by squaring the partial root, multiplying by three, and adding two ciphers. The addition to be made to this partial root is something less than the quotient of the second dividend and the partial divisor. It has to be found by trial and is here called the "trial addition to the root." The partial divisor is increased by—

(*a*) Three times the product of the partial root (with one cipher added) and the trial addition to the root ;

(*b*) The square of the trial addition to the root.

This gives the second divisor, which is then multiplied by the trial addition to the root. If the product is greater than the dividend, the trial addition is too large.

If the dividend exceeds the product by an amount greater than the divisor, the trial addition is too small.

When a trial addition has been found to avoid these extremes, the next period from the original number is added to the remainder and the work proceeds as before.

The simplest method of extracting roots is by means of logarithms, and these will be explained later.

The fourth root of a number can be obtained by finding its square root, and then extracting the square root of the answer.

Using Tables of Roots.—Most engineering pocket-books include tables of roots and powers of numbers up to 100 or 200, and the use of these can be extended to apply to larger numbers, by applying the rule that *the square of a composite number* (as distinct from a *prime* number) *is equal to the square of its two factors multiplied together*. For example, suppose the table of powers extends only to 100, and it is desired to find the square of 123. Mentally divide 123 by 3 ; the answer is 41. Find the square of 41 from the table and multiply the answer by the square of 3=9. The rule to remember is : *Divide the number to be squared by the smallest number or factor as will bring it within the compass of the table of powers,* thus taking advantage of the highest square in relation to the number which the table will yield.

The same rule applies when extracting roots, for *the square root, cube root, or any other root of a number is equal to the root of one of its factors multiplied by the same root of the other factor.*

Example : To find the cube root of 512 from a table of roots which does not go beyond 100, divide the number by some smaller cube number to bring it within the compass of the table. In this case we can divide by 8 (2 cubed), the quotient being 64. From the table we note that the cube root of 64 is 4, and multiplying this by the cube root of 8 (2) we obtain $4\times2=8$ as the cube root of 512.

CHAPTER VII

Some Short Cuts

Practical men use a vast number of short cuts in calculations. A few of the more useful are given.

To multiply by 5, add nought to the number to be multiplied and divide by 2.

To multiply by 25, add two noughts and divide by 4.

To multiply by 125, add three noughts and divide by 8.

To divide by 5, multiply by 2 and divide by 10.

To divide by 25, multiply by 4 and divide by 100.

To divide by 125, multiply by 8 and divide by 1000.

Division and multiplication by contracted methods have already been given.

A convenient method of squaring a number is to multiply the number *plus* the unit figure by the number *less* the unit figure, and add the square of the unit. Example: Square 92.

$$92+2=94;\quad 92-2=90$$
$$94\times 90=8460$$
$$2\text{ squared}=4=\frac{4}{8464}$$

A quick way of multiplying by 9 is to add a nought to the number to be multiplied and then subtract the number.

Example: $321\times 9=3210-321=2889$.

Similarly, when multiplying by 99, add two noughts and subtract the number; when multiplying by 999 add three noughts, and so on.

Contracted Multiplication.—There is a contracted system of multiplication which saves considerable time when results are only required to be accurate to the first one or two places of decimals. I give an example, showing the usual and contracted methods : Multiply ·007435 by 6·325.

Normal Method.	Contracted Method.
7435	7435
6325	5236
37175	44610
14870	2230⋮5
22305	148⋮70
44610	37⋮175
·047026375	047025

It will be noted that in the contracted method the figures used in the multiplier are reversed, and the rows of figures resulting from each multiplication are arranged one place to the right of the previous result of multiplication. Those figures to the right of the dotted line would not, of course, be written down in practice. They are ignored, and are merely included here to indicate the method of working. In practice, therefore, the first figure is ignored, although any number to be carried is added in the usual way. The working would thus appear :

```
 7435
 5236
 ----
44610
 2230
  148
   37
 ----
·047025
```

Contracted Division.—Similarly, it is possible to contract the process of division and one example will suffice : Divide ·031625 by 3·125. Ordinary method :

```
3125)31·625(·01012
     31 25
     -----
      3750
      3125
      ----
       6250
       6250
```

By observation 3 divides into 31 (the first two significant figures of the dividend) 10 times. Multiply the divisor by 10, producing 3125. Subtract this from the dividend, leaving 375. Now drop the 5 from the divisor, and it will be seen that 312 will divide once into the remainder 375. (If in other examples this shortened divisor divides into the remainder more than once—say 5 times—then the figure dropped is multiplied by 5 and the first figure ignored, whilst the second is carried on.) Deduct 312 from 375, leaving 63. Now drop another figure from the divisor, leaving 31, which divides twice into 63. We thus have :

1st division by observation			10
2nd ,, ,, contraction			1
3rd ,, ,, ,,			2
			·01012

The position of the decimal point is, of course, fixed in the manner already explained, and the result is only approximate.

To ascertain whether any number is divisible by 3, add up the digits, and if they are divisible by 3 then the number itself is. For example, take the number 54321 ; these digits add up up to 15, and as 15 is divisible by 3, 54321 is divisible by 3. Another example : the sum of the digits of the number 9786354531 is 51, and as 51 is divisible by 3, we know that 3 is a factor of 9786354531. There is no other factor except 9 to which this rule applies.

If the last two figures of any number are divisible by 4 then that number itself is divisible by 4. For example, take 49732 ; the last two figures (32) are divisible by 4.

To ascertain whether a number is divisible by 11, add up the first, third, and fifth, etc., digits, and the second, fourth, sixth, etc., digits. Subtract the smaller number from the larger. If the answer is either 0 or 11, then the number is divisible by 11. For example, 152273. Adding every other number, $1+2+7=10$, and $5+2+3=10$; $10-10=0$. Therefore 152273 is divisible by 11.

Another example : 1647382 ; $1+4+3+2=10$; $6+7+8=21$; $21-10=11$.

CHAPTER VIII

Logarithms

Logarithms are one of the most important of the labour-saving methods of calculation. It is necessary to understand the use of logarithmic tables, for they enable calculations to be made which are almost impossible by any other method. By means of logarithms (logs. for short), we are able to carry out the operations of multiplication, division, involution, and evolution (not, be it noted, of addition or subtraction). A knowledge of logarithms is quite essential before the slide rule (a most useful instrument) can be mastered. The slide rule itself considerably augments the rapid calculation which logarithms provide. I shall deal here first with *common logarithms*, leaving *Napierian logarithms* or *Hyperbolic logarithms* until later. Common logarithms are those calculated to base 10.

In logarithms, *the processes of multiplication and division are converted into those of addition and subtraction.* Thus, to multiply numbers together, their logarithms are *added*, and to divide them their logarithms are *subtracted*. Let us first define what a logarithm is, and then obtain an idea of how logarithms are applied.

The logarithm of a number, to a given base, is the index of the power to which the base must be raised to produce the aforesaid number. Here is a table of the number 3 raised to various powers :

$3^1=3$	$3^7=2187$
$3^2=9$	$3^8=6561$
$3^3=27$	$3^9=19683$
$3^4=81$	$3^{10}=59049$
$3^5=243$	$3^{11}=177147$
$3^6=729$	$3^{12}=531441$

Suppose we wish to multiply 27 by 19683. We can proceed by the lengthy way as follows :

$$\begin{array}{r} 19683 \\ 27 \\ \hline 137781 \\ 393660 \\ \hline 531441 \end{array}$$

By means of the table of logarithms, however, we look up the *logarithm* of the two numbers, add the two logarithms together, and then consult the table to ascertain which number corresponds to the logarithm so obtained. Thus, from the table we see that—

$$27 = 3^3, \text{ and } 19683 = 3^9$$

The two logarithms are therefore 3 and 9 which, *added* together, equal 12.

Consulting the table we see that $3^{12} = 531{,}441$.

Observe that we have not *multiplied* the quantities to obtain the answer.

The table above is calculated to the *base* 3. In other words, it contains the index of the *power* (1, 2, 3, . . . 12) to which the *base* (3) must be raised to produce the numbers (3 to 531,441).

If we wished to *divide* two of the numbers in the table we should *subtract* their logarithms. For example, $19683 \div 27$. By ordinary methods this is found to be 729.

But the logarithm of 27 (from the table) is seen to be 3, and the logarithm of 19683 is 9. Subtracting the logarithms, $9 - 3 = 6$; the answer therefore to $19683 \div 27$ is $3^9 - 3^3 = 3^{9-3} = 3^6$. From the table we see that $3^6 = 729$, the answer.

These are simple examples of the *principle* of logarithms.

Common logarithms are, however, *calculated to the base* 10. It is obvious that in multiplication and division by logarithms the numbers to be multiplied or divided will seldom be an exact power of 10. If they were, logarithms would not be needed, for we should merely add noughts to the multiplicand according to the number of digits in the multiplier. Thus, $10{,}000 \times 100 = 1{,}000{,}000$.

So let us make another table of powers of 10.

$$1{,}000{,}000=10^6$$
$$100{,}000=10^5$$
$$10{,}000=10^4$$
$$1{,}000=10^3$$
$$100=10^2$$
$$10=10^1$$
$$1=10^0$$
$$\cdot 1=\frac{1}{10}=\frac{1}{10^1}=10^{-1}$$
$$\cdot 01=\frac{1}{100}=\frac{1}{10^2}=10^{-2}$$
$$\cdot 001=\frac{1}{1000}=\frac{1}{10^3}=10^{-3}$$
$$\cdot 0001=\frac{1}{10{,}000}=\frac{1}{10^4}=10^{-4}$$
$$\cdot 00001=\frac{1}{100{,}000}=\frac{1}{10^5}=10^{-5}$$
$$\cdot 000001=\frac{1}{1{,}000{,}000}=\frac{1}{10^6}=10^{-6}$$

The Characteristic and the Mantissa.—It is obvious that any number between, say, 1,000,000 and 100,000 would have an *index* or *characteristic* between 5 and 6, and it is the decimal part of the index for which we require tables of *logarithms*. This decimal part of the index or characteristic is called the *mantissa*. Thus the logarithm of a number consists of two parts—the *index* or *characteristic*, and the *mantissa*.

Now the characteristic (to base 10) of any number can quite easily be found by observation.

The characteristic of any number greater than unity is always 1 *less than the number of digits in the number*.

Thus in the table above :

1,000,000 has 7	digits, and	the	characteristic is	6
100,000 has 6	,,	,,	,,	5
10,000 has 5	,,	,,	,,	4
1,000 has 4	,,	,,	,,	3
100 has 3	,,	,,	,,	2
10 has 2	,,	,,	,,	1
1 has 1	,,	,,	,,	0

The characteristic of any number less than unity is greater by 1 than the number of noughts which follow the decimal point.

The index or characteristic of a number less than unity is also negative; this is indicated by placing the negative sign or bar over the figure, as in the table.

Here are a few examples:

Index or	characteristic	of	547·062	equals	2
,,	,,	,,	47·062	,,	1
,,	,,	,,	7·062	,,	0
,,	,,	,,	·7062	,,	$\bar{1}$
,,	,,	,,	·07062	,,	$\bar{2}$
,,	,,	,,	·007062	,,	$\bar{3}$
,,	,,	,,	·0007062	,,	$\bar{4}$

The characteristic of 100 is 2, and of ·01, $\bar{2}$. The characteristic of any number over 100 but less than 1000 will also be 2, *plus* the mantissa found from a table of logarithms. The characteristics of any decimal from ·1 to ·9 will always be $\bar{1}$, the characteristic of any decimal from ·01 to ·09 will be $\bar{2}$, the characteristic of any decimal from ·001 to ·009 will be $\bar{3}$, and so on.

The numbers, whether whole or decimal, or whether consisting of a whole number and a decimal, are treated as whole numbers for the purpose of extracting the mantissa from the tables.

The characteristic of a number which is a decimal only, and has no whole numbers before the decimal point, is always *negative*. The characteristic of a number consisting of whole numbers and a decimal will always be *positive*, and the decimal part in this latter case is ignored when determining the characteristic by inspection. Thus, in the number 4890·375, we ignore the decimal ·375, and as the whole number contains 4 digits the characteristic will be 3. We must, however, take into account both the whole number and the decimal when finding the mantissa from the tables. We, in fact, proceed as if the decimal point did not exist.

It is important to remember that the mantissa is always positive; and the mantissa of a number, irrespective of whether it is a decimal or contains a whole number and a decimal, contains the same digits. It is only the characteristic which varies, according to the number. For example, the logarithm of 101 is 2·0043 ; the logarithm of 1·01 is 0·0043, of ·101 it is $\bar{1}$·0043, and so on. The logarithm of all numbers from 1 to 9·999 . . . will consist of decimals only.

Use of Log. Tables.—Some tables of logarithms are calculated to four places of decimals, and these are sufficiently accurate for most purposes. By using such tables we are only able to extract the logarithms of the first four figures of numbers ; hence, if we wish to extract the logarithm of 605·125 we must abbreviate the number by the method described in Chapter IV. In this case we should extract the logarithm of 605·1.

If the number, however, was, say, 605·175, we should (as 7 is more than 5) call it 605·2. Again, if the expression were 605·045 we should write it as 605·0, because ·045 is less than ·05 and is therefore nearer to ·0 than to ·1.

Find the logarithm of 126·5. The characteristic is 2, and we look up the first two figures of the number (12) in the first column of the table of logarithms. We next run our eye along that line and read beneath the third figure of the number (6), 1004 ; continuing, under 5 in what is known as the *difference column,* we read 17. We add 17 to 1004, making 1021.

Therefore, 126 equals $10^{2\cdot1021}$.

Find the logarithm of 146·0. Answer, 2·1644.
Find the logarithm of 17·53. Answer, 1·2437.
Find the logarithm of 1·399. Answer, 0·1459.
Find the logarithm of ·1621. Answer, $\bar{1}$·2098.
Find the logarithm of ·01788. Answer, $\bar{2}$·2524.

The logarithms represent the power to which the base 10 must be raised to produce the number.

This description and these examples are given to indicate how we extract the logarithms of numbers from tables, for the object of carrying out the processes of multiplication, division, involution, and evolution, when we have large and unwieldy quantities with which to deal, or when we have a large number of calculations to make. We have seen that in order to multiply in logarithms, we merely add the logarithms.

I will now give some examples : Multiply 165·3 by 144·6.

The logarithm of 165·3 is found to be 2·2183, and the logarithm of 144·6 is 2·1602. Add these together:

$$\begin{array}{r} 2{\cdot}2183 \\ 2{\cdot}1602 \\ \hline 4{\cdot}3785 \end{array}$$

We now look up the table of antilogarithms (ignoring the characteristic) and we find that ·3785 corresponds to 2391. We now have to fix the position of the decimal point. We have seen that the characteristic is always 1 less than the number of digits. As the characteristic is 4 there will be 5 figures before the decimal point in the answer. Therefore, the answer is 23910·0, which is approximately correct, because if we multiply 165·3 by 144·6 in the ordinary way we shall obtain 23902·38.

Multiply 789·36 by 284·87. As we are using four-figure tables we must *contract* each of the two quantities to four figures. Thus 789·36 becomes 789·4, and 284·87 becomes 284·9.

The log. of 789·4 is 2·8973
The log. of 284·9 is 2·4547

Adding 5·3520
Antilog. of ·352=2249.

As the characteristic is 5 there will be six places before the decimal point, and the answer therefore is 224,900·0, which is approximately correct.

In the two preceding examples the characteristic is positive. In cases where the characteristic of one quantity is positive and in the other negative, slightly different treatment is necessary. For example: Multiply 37·65 by ·0135.

Log. 37·65=1·5758
and log. ·0135=$\bar{2}$·1303

Adding $\bar{1}$·7061

Here it will be seen that the positive characteristic added to the negative produces $\bar{1}$.

Antilog. ·7061=5083.

As the characteristic is $\bar{1}$ there will be no ciphers before the decimal point.

Answer: ·5083.

Similarly, when adding two negative characteristics. Suppose the two logs. are $\bar{2}\cdot3064$ and $\bar{5}\cdot0913$.

$$\begin{array}{rr} \text{Adding} & \bar{2}\cdot3064 \\ & \bar{5}\cdot0913 \\ \hline & \bar{7}\cdot3977 \end{array}$$

Antilog.$=2499$.

Answer: $\cdot000000249$.

The treatment is different in cases where the addition of the two mantissæ provides a " carry over."

For example, in adding the following two logarithms :

$$\begin{array}{rr} & 2\cdot3164 \\ & \bar{3}\cdot9112 \\ \hline \text{Adding} & \cdot2276 \end{array}$$

Here it will be seen that, in adding the 9 and 3 of the mantissa, there is a positive 1 to carry to the left of the decimal point. (We have already seen that the mantissa is always positive.)

Adding this 1 to the positive 2, we obtain 3, which cancels out the negative $\bar{3}$. Another example : Add together log. $\bar{3}\cdot1654$ and $\bar{1}\cdot9733$.

$$\begin{array}{r} \bar{3}\cdot1654 \\ \bar{1}\cdot9733 \\ \hline \bar{3}\cdot1387 \end{array}$$

The two negative characteristics added together equal $\bar{4}$, and as there is a positive 1 carried from the addition of the mantissæ, the characteristic of the addition becomes $\bar{3}$, because $\bar{4}+1=\bar{3}$.

Division by Logarithms.—*Division by logarithms is effected by subtracting the logarithm of the divisor from that of the dividend, the result of the subtraction being the logarithm of the quotient.*

Example : Divide 37·65 by 19·01.

$$\begin{array}{lr} \text{Log. } 37\cdot65 & =1\cdot5758 \\ \text{Log. } 19\cdot01 & =1\cdot279 \\ \hline \text{Subtracting:} & 0\cdot2968 \end{array}$$

Antilog.$=1981$.

As the characteristic is 0, there will be one digit (0+1) before the decimal point, and the answer is 1·981.

Divide 5·065 by ·0015.

$$\begin{aligned} \text{Log. } 5{\cdot}065 &= 0{\cdot}7046 \\ \text{Log. } {\cdot}0015 &= \bar{3}{\cdot}1761 \\ \hline \text{Subtracting : } & 3{\cdot}5285 \end{aligned}$$

Antilog. of ·5285=3377.

As the characteristic is 3, there will be four figures before the decimal point. Note that the characteristic of the log. being subtracted is changed from negative to positive.

Answer: 3377·0.

In subtraction, change the sign of the characteristic being subtracted and add.

Involution.—Logarithms may be used to perform the functions of involution and evolution. The rule is : *To evaluate the power of a number (as* $7{\cdot}5^3$*), multiply the logarithm of the number by its index, and the result is the logarithm of the number required.* Consult the table of antilogarithms for the number corresponding.

Example : Find the number which equals $7{\cdot}5^3$.

$$\begin{aligned} \text{Log. } 7{\cdot}5 &= 0{\cdot}8751 \\ \text{Multiply by index 3} & \quad\quad 3 \\ \hline & 2{\cdot}6253 \end{aligned}$$

Antilog. ·6253= 4220.

Characteristic is 2, therefore there will be three figures before the decimal point.

Answer, 422·0 (actually 421·875).

Evaluate $\cdot0735^5$.

$$\begin{aligned} \text{Log. } {\cdot}0735 &= \bar{2}{\cdot}8663 \\ \text{Multiply by index} & \quad\quad 5 \\ \hline & \bar{6}{\cdot}3315 \end{aligned}$$

Antilog. ·3315= 2145.

Characteristic is $\bar{6}$, so there will be five ciphers after the decimal point. Answer, ·000002145.

Remember that the mantissa is always positive, so in multiplying, the carry over from the decimal part will be positive and must be *subtracted* if the characteristic is negative. In the above example there was 4 to carry; $5\times\bar{2}=\overline{10}$, and subtracting the 4 leaves $\bar{6}$.

When the index of a number consists of several figures, and the number is less than unity (the characteristic thus being negative) the whole logarithm must be converted into a negative number before multiplying by the index.

Example: Evaluate $\cdot735^{-3\cdot75}$.

$$\text{Log. } \cdot735=\bar{1}\cdot8663$$
$$=-1+\cdot8663$$
$$=-\cdot1337 \text{ (}\cdot8663\text{ has been subtracted from 1)}$$

```
                    — ·1337
Multiply by index   —3·75
                    ------
                     6685
                    9359
                   4011
                   -------
                   ·501375
```

Antilog. $\cdot5014=3173$.

Answer: 3·173.

Remember: *When negative signs are multiplied together the result is positive; but when the logarithm is positive and the index negative, the product is negative. In the latter case the mantissa must be made positive before referring to the table of antilogarithms.*

Example: Evaluate $(7\cdot5)^{-1\cdot5}$

$$\text{Log. } 7\cdot5=0\cdot8751$$
$$-1\cdot5\times0\cdot8751=-1\cdot31265.$$

Here the mantissa is negative, and it is made positive in the following way:

$$-\cdot31265=\bar{1}\cdot68735$$
$$\text{Therefore } -1\cdot31265=\bar{2}\cdot68735$$
$$\text{Antilog. } \cdot6874=4868$$

Answer: ·0468.

The expression $(7{\cdot}5)^{-1{\cdot}5}$ can, of course, also be written $\frac{1}{7{\cdot}5^{1{\cdot}5}}$, and by working it out in this form verification of the above result can be obtained.

Evolution.—*The root of a number can be obtained by dividing the logarithm of the number by the required root.*

Example : Evaluate $\sqrt[3]{28{\cdot}06}$

$$\text{Log. } 28{\cdot}06 = 1{\cdot}4481$$
$$\text{Divide by the root } 3 = {\cdot}4827$$
$$\text{Antilog. } {\cdot}4827 = 3{\cdot}039$$

Answer : 3·039.

The beginner finds no difficulty in evolution by logs. when characteristic and mantissa are both positive. When the characteristic is negative it must be adjusted so that the logarithm is exactly divisible by the root.

Example : Find the cube root of ·625.

$$\text{Log. } {\cdot}625 = \bar{1}{\cdot}7959.$$

The characteristic is negative and the mantissa positive in this case. Therefore we must add the smallest number to the characteristic to make it divisible by 3. We, hence, add $\bar{2}$ to $\bar{1}$, making $\bar{3}$. To preserve balance we must add $+2$ to the mantissa ; thus $\bar{1}{\cdot}7959$ now becomes $\bar{3}+2{\cdot}7959$.

Dividing by 3 the expression becomes—

$$\bar{1}{\cdot}93196.$$

$$\text{Antilog. } {\cdot}932 = 8551.$$

Answer : ·8551.

The adjustment of the characteristic is, of course, performed mentally once the principle is mastered.

It is important to remember that, when dividing a logarithm by a number which is greater than the first figure in the mantissa, a cipher must be added. For example : Find the seventh root of 4.

$$\text{Log. } 4 = {\cdot}6021$$
$$\frac{1}{7}({\cdot}6021) = {\cdot}08601$$
$$\text{Antilog. } {\cdot}08601 = 1219$$

Answer : 1·219.

Here it will be seen that 7 will not divide into 6, so a cipher is added ; 7 divided into 60 gives 8, and so on.

Find the 5th root of ·009, or $\cdot 009^{\frac{1}{5}}$.

$$\text{Log. } \cdot 009 = \bar{3}\cdot 9542$$
$$\frac{1}{5}(\bar{3}\cdot 9542) = \frac{1}{5}(\bar{5} + 2\cdot 9542)$$
$$= \bar{1}\cdot 5908$$
$$\text{Antilog. } \cdot 5908 = 3897$$

Answer: ·3897.

The principle of logarithms was discovered by Napier, but Napierian logarithms (sometimes termed hyperbolic or natural logarithms, as distinct from *common* logarithms calculated to base 10), are calculated to a base which is the sum of the series :

$$1 + \frac{1}{2} + \frac{1}{2\times 3} + \frac{1}{2\times 3\times 4} + \frac{1}{2\times 3\times 4\times 5} \ \ldots .$$

The sum of this series to seven places of decimals is 2·7182818 denoted by the symbol e. To convert common into Napierian or hyperbolic logarithms, multiply by 2·3026 or, more accurately, 2·30258509.

To convert Napierian into common logarithms, multiply by ·4343. . . .

To convert logarithms calculated to base a to logarithms of base b proceed algebraically, letting x=log. of number calculated to base a, and y=log. of number calculated to base b.

$$\text{Let } N = \text{the number}$$
$$\text{Then } N = a^x$$
$$\text{and } N = b^y$$
$$\text{Therefore } a^x = b^y$$
$$\text{and } b = a^{\frac{x}{y}}$$
$$\frac{x}{y} = \text{log. }_a b$$
$$\frac{y}{x} = \frac{1}{\text{log. }_a b}$$

Common logarithms were calculated in this way from Napierian logarithms.

CHAPTER IX

The Metric System

The French or metric system used on the Continent, and which is gradually being introduced in this country, is a great improvement on the British systems of measurement. It is a decimal system in which 10 is the unit. The unit of length is the *metre,* which is one 10,000,000th part of the length of the meridian on the north and south line extending from the equator from one of the poles. Later it was found that this determination was wrong, and nowadays the metre is defined as the distance at 0 degree cent. between two marks on a bar of platinum. This bar is kept at Paris, and is known as the *Metre des Archives.* The unit of square measure is the *are,* the unit of weight the *gramme,* and the unit of capacity the *litre.*

Prefixes of these units denote multiples and submultiples.

Mega means a million times.
Myria means ten thousand times.
Kilo means a thousand times.
Hecto means a hundred times.
Deca (or *Deka*) means ten times.
deci means a tenth part of.
centi means a hundredth part of.
milli means a thousandth part of.
micro means a millionth part of.

Capitals are used for the multiples, and small letters for the sub-multiples.

Length

10 millimetres (mm.)	=1 centimetre (cm.).
10 centimetres	=1 decimetre (dm.).
10 decimetres	=1 Metre (m.).
10 Metres	=1 Decametre (Dm.) (sometimes Dekametre).
10 Decametres	=1 Hectometre (Hm.).
10 Hectometres	=1 Kilometre (Km.).
10 Kilometres	=1 Myriametre (Mm.).

Square Measure

100 square millimetres (mm.²)	=1 square centimetre (cm.²)
100 square centimetres	=1 square decimetre dm.²)
100 square decimetres	=1 square metre (m.²)
100 square metres	=1 Are (ar.)
100 ares	=1 hectare (har.)
100 hectares	=1 square Kilometre (Km.²)

Weight

10 milligrams (mg.)	=1 centigram (cg.)
10 centigrams	=1 decigram (dg.)
10 decigrams	=1 gram (g.)
10 grams	=1 Decagram (Dg.)
10 Decagrams	=1 Hectogram (Hg.)
10 Hectograms	=1 Kilogram (Kg.)
10 Kilograms	=1 Myriagram
10 Myriagrams	=1 quintal
10 quintals	=1 Millier or tonneau

Capacity

1 litre	=1 cubic Decimetre.
10 litres	=1 Decalitre.
10 decalitres	=1 Hectolitre.
10 hectolitres	=1 Kilolitre.

A Litre is a Kilogramme of water at 4 deg. C.

The simplicity of being able to multiply and divide in the metric system is apparent, for it is only a question of moving the decimal point. Thus, the metre is 39·37 inches approximately. A millimetre will be a thousandth part of that, or ·03937, and a kilometre will be one thousand times 39·37, or 39,370·000 inches.

METRIC CONVERSION FACTORS

Equivalents of Imperial and Metric Weights and Measures

Linear Measure

Imperial

1 Inch	=	25·400 Millimetres.
1 Foot	=	0·30480 Metre.
1 Yard	=	**0·914399 Metre.**
1 Fathom	=	1·8288 Metres.
1 Pole	=	5·0292 ,,
1 Chain	=	20·1168 ,,
1 Furlong	=	201·168 ,,
1 Mile	=	1·6093 Kilometres.

Metric

1 Millimetre (mm.) ($\frac{1}{1000}$ m.)	=	0·03937 Inch.
1 Centimetre ($\frac{1}{100}$ m.)	=	0·3937 ,,
1 Decimetre ($\frac{1}{10}$ m.)	=	3·937 Inches.
1 Metre (m.)	=	**39·370113 Inches.** **3·280843 Feet.** **1·0936143 Yards.**
1 Decametre (10 m.)	=	10·936 Yards.
1 Kilometre (1,000 m.)	=	0·62137 Mile.

Square Measure

Imperial

1 Square Inch	=	6·4516 Square Centimetres.
1 Square Foot	=	9·2903 Square Decimetres.
1 Square Yard	=	0·836126 Square Metre.
1 Rood	=	10·117 Ares.
1 Acre	=	0·40468 Hectare.
1 Square Mile	=	259·000 Hectares.

Metric

1 Square Centimetre	=	0·15500 Square Inch.
1 Square Decimetre	=	15·500 Square Inches.
1 Square Metre	=	10·7639 Square Feet. 1·1960 Square Yards.
1 Are	=	119·60 Square Yards.
1 Hectare	=	2·4711 Acres.

Cubic Measurement

Imperial

1 Cubic Inch	=	16·387 Cubic Centimetres.
1 Cubic Foot	=	0·028317 Cubic Metre.
1 Cubic Yard	=	0·764553 ,, ,,

Metric

1 Cubic Centimetre	=	0·0610 Cubic Inches.
1 Cubic Decimetre (c.d.)	=	61·024 ,, ,,
Cubic Metre	=	35·3148 Cubic Feet. 1·307954 Cubic Yards.

Measure of Capacity

Imperial

1 Pint	=0·568 Litre.
1 Quart	=1·136 Litres.
1 Gallon	**=4·5459631 Litre.**

Metric

1 Centilitre ($\frac{1}{100}$ litre)	=0·070 Gill.
1 Decilitre ($\frac{1}{10}$ litre)	=0·176 Pint.
1 Litre	**=1·75980 Pints.**

Weight

Avoirdupois

1 Grain	=	0·0648 Gramme.
1 Dram	=	1·772 Grammes.
1 Ounce	=	28·350 ,,
1 Pound (7000 grains)	=	**0·45359243 Kilogramme.**
1 Hundredweight	=	50·80 Kilogramme. 0·5080 Quintal.
1 Ton	=	1·0160 Tonnes of 1016 Kilogrammes.
1 Grain (Troy)	=	0·0648 Gramme.
1 Troy Ounce	=	31·1035 Grammes.
1 Milligramme ($\frac{1}{1000}$ grm.)	=	0·015 Grain.
1 Centigramme ($\frac{1}{100}$ grm.)	=	0·154 ,,
1 Gramme (1 grm.)	=	15·432 Grains.
1 Kilogramme (1000 grm.)	=	**2·2046223 Lb. or 15,432·3564 grains.**
1 Quintal (100 kilog.)	=	1·968 Cwt.
1 Tonne (1000 kilog.)	=	0·9842 Ton.
1 Gramme (1 grm.)	=	0·03215 Oz. Troy. 15·432 Grains.

METRIC TO ENGLISH CONVERSION FACTORS

To Convert

Millimetres to inches	×0·03937 or ÷25·4.
Centimetres to inches	×0·3937 or ÷2·54.
Metres to inches	×39·37.
Metres to feet	×3·281

Metres to yards	×1·094.
Metres per second to feet per minute	×197.
Kilometres to miles	×0·6214 or ÷1·6093.
Kilometres to feet	×3280·8693.
Square millimetres to square inches	×0·00155 or ÷645·1.
Square centimetres to square inches	×0·155 or ÷6·451.
Square metres to square feet	×10·764.
Square metres to square yards	×1·2.
Square kilometres to acres	×247·1.
Hectares to acres	×2·471.
Cubic centimetres to cubic inches	×0·06 or ÷16·383.
Cubic metres to cubic feet	×35·315.
Cubic metres to cubic yards	×1·308.
Cubic metres to gallons (231 cu. in.)	×264·2.
Litres to cubic inches	×61·022.
Litres to gallons	×0·21998 or ÷4·545.
Hectolitres to cubic feet	×3·531.
Hectolitres to bushels (2,150·42 cu. in.)	×2·84.
Hectolitres to cubic yards	×0·131.
Grammes to ounces (avoirdupois)	×0·035 or ÷28·35.
Grammes per cubic centimetre to pounds per cubic inch	÷27·7.
Joules to foot-pounds	×0·7373.
Kilogrammes to ounces	×35·3.
Kilogrammes to pounds	×2·2046.
Kilogrammes to tons	×0·001.
Kilogrammes per square centimetre to pounds per square inch	×14·223.

Kilos per lineal metre	×0·672	=pounds per lineal foot.
Kilos per lineal metre	×2·016	=pounds per lineal yard.
Kilos per lineal metre	×0·0003	=tons per lineal foot.
Kilos per lineal metre	×0·0009	=tons per lineal yard.
Kilos per kilometre	×3·548	=pounds per mile.
Kilos per square centimetre	×14·223	=pounds per square inch.
Kilos per square millimetre	×0·635	=tons per square inch.
Kilos per square metre	×0·2048	=pounds per square foot.
Tonnes per square metre	×0·0914	=tons per square foot.
Tonnes per square metre	×0·823	=tons per square yard.
Kilos per cubic metre	×1·686	=pounds per cubic yard.
Kilos per cubic metre	×0·0624	=pounds per cubic foot.
Tonnes per cubic metre	×0·752	=tons per cubic yard.
Grammes per litre	×70·12	=grains per gallon.

Kilos per litre	×10·438	=pounds per gallon.
Litres per square metre	×0·0204	=gallons per square foot.
Kilogrammetres	×7·233	=foot-pounds.
Kilogrammetres	×0·0387	=inch-tons.
Tonne-metres	×3·23	=foot-tons.
Force de cheval	×0·9863	=horse-power.
Kilos per cheval	×2·235	=pounds per H.P.
Square metres per cheval	×10·913	=square feet per H.P.
Cubic metres per cheval	×35·806	=cubic feet per H.P.
Calories	×3·968	=heat units.
Calories per square metre	×0·369	=heat units per square foot.

English to Metric Conversion Factors

1 inch	=2·54 centimetres, or 25·4 millimetres.
1 foot	=30·4799 centimetres, 304·799 millimetres, or 0·3047 metre.
1 yard	=0·914399 metre.
1 mile	=1·6093 kilometres=5280 feet.
1 millimetre	=0·03937 inch.
1 centimetre	=0·3937 inch.
1 decimetre	=3·937 inches.
1 Hectometre	=0·0621 mile.
1 Metre	=39·370113 inches. 3·28084 feet. 1·093614 yards.
1 Kilometre	=0·62137 mile.
1 Decametre (10 metres)	=10·936 yards.

Compound Conversion Factors

Pounds per lineal foot	×1·488	=kilos per lineal metre.
Pounds per lineal yard	×0·496	=kilos per lineal metre.
Tons per lineal foot	×3333·33	=kilos per lineal metre.
Tons per lineal yard	×1111·11	=kilos per lineal metre.
Pounds per mile	×0·2818	=kilos per kilometre.
Pounds per square inch	×0·0703	=kilos per square centimetre.
Tons per square inch	×1·575	=kilos per square millimetre.
Pounds per square foot	×4·883	=kilos per square metre.
Tons per square foot	×10·936	=tonnes per square metre.

Tons per square yard	×1·215	=tonnes per square metre.
Pounds per cubic yard	×0·5933	=kilos per cubic metre.
Pounds per cubic foot	×16·020	=kilos per cubic metre.
Tons per cubic yard	×1·329	=tonnes per cubic metre.
Grains per gallon	×0·01426	=grammes per litre.
Pounds per gallon	×0·09983	=kilos per litre.
Gallons per square foot	×48·905	=litres per square metre.
Inch-tons	×25·8	=kilogrammetres.
Foot-pounds	×0·1382	=kilogrammetres.
Foot-tons	×0·309	=tonne-metres.
Horse-power	×1·0139	=force de cheval.
Pounds per H.P.	×0·477	=kilos per cheval.

British Unit of Length.—The Imperial Standard Yard is the distance between the centres of two gold plugs or pins in the bronze bar used for determining the Imperial Standard Yard, measured when the bar is at a temperature of 62 deg. F. It is supported on bronze rollers placed under it to avoid flexure of the bar and to facilitate its free expansion and contraction from variations of temperature. The Imperial Standard is a solid square bar 38 in. long and 1 in. square in section. Near to each end a cylindrical hole is sunk exactly 36 in. from centre to centre. At the bottom of each hole is inserted a gold plug about $\frac{1}{10}$th of an inch in diameter, and upon the surface of the pins are cut three fine lines at intervals of $\frac{1}{100}$th of an inch, transverse to the axis of the bar, and two lines at nearly the same interval parallel to the axis of the bar. These are the datums from which the measurement is taken. There is no need to give here a complete list of British weights and measures.

A complete table of English weights and measures is given on pages 207 to 210.

CHAPTER X

Progression: Arithmetical, Geometrical, and Harmonical

A series of numbers which increase or decrease by a constant difference is known as an *Arithmetical Progression.* Thus, 1, 4, 7, 10, 13, 16, . . . is an arithmetical progression, the common difference being +3. The series 12, 10, 8, 6, 4, 2 is also an arithmetical progression, the difference here being —2. We can reduce this to a formula so that when we wish to know the *sum* of such a series we do not need to add each term of the series together separately. Of course, in a simple series like that given the addition sum could be performed mentally, but in a complicated series, where each term of the series is a large number or consists of decimals, it is more convenient to use the formula.

Let a=the first term of the series, z the last term, n the number of terms, d the constant difference, and S the sum of all the terms. We can now arrive at a series of formulæ from which we can find the first term of a series from a total representing the sum of the series, or we can find the last term, knowing the first and the number of terms and the difference; we can find the number of terms from the sum, the difference, or, in fact, any function of the series. Here are the formulæ:

$$a=z-d(n-1). \quad a=\frac{2S}{n}-z. \quad a=\frac{S}{n}-\frac{d}{2}(n-1)$$

$$z=a+d(n-1).$$

$$z=\frac{2S}{n}-a. \quad z=\frac{S}{n}+\frac{d}{2}(n-1). \quad n=\frac{z-a}{d}+1$$

$$n=\frac{2S}{a+z}$$

$$d=\frac{z-a}{n-1}. \quad d=\frac{(z+a)\ (z-a)}{2S-a-z}. \quad d=\frac{2(S-an)}{n(n-1)}$$

$$d=\frac{2(zn-S)}{n(n-1)}$$

$$S=\frac{n(a+z)}{2}.\quad S=\frac{(a+z)\,(z+d-a)}{2d}.\quad S=n[a+\frac{d}{2}(n-1)]$$

$$S=n[z-\frac{d}{2}(n-1)].\quad a=\frac{d}{2}\pm\sqrt{\left(z+\frac{d}{2}\right)^2-2dS}$$

$$z=\frac{d}{2}\pm\sqrt{\left(a-\frac{d}{2}\right)^2+2dS}$$

$$n=\frac{1}{2}-\frac{a}{d}\pm\sqrt{\left(\frac{1}{2}-\frac{a}{d}\right)^2+\frac{2S}{d}}$$

$$n=\frac{1}{2}+\frac{z}{d}\pm\sqrt{\left(\frac{1}{2}+\frac{z}{d}\right)^2-\frac{2S}{d}}$$

When the series is decreasing make the first term $=z$, and the last term a.

The Arithmetical Mean of two quantities, A and $B=\frac{A+B}{2}$

Examples in Arithmetical Progression.—I now give examples. I have presumed the same series throughout so that the reader may more easily follow the applications, 5 being the first term of the series described, the last term 125, and the constant difference 5, the sum being 1625. It is important to select the correct formula according to whether it is the sum, the last term, or the difference which is to be found. Thus, if the first term of a series is to be found when the last term and the difference is known, a formula must be selected where a is the first unknown quantity.

Example.—The first term of a series in arithmetical progression is 5, and the last term 125, the constant difference being 5. Find the number of terms in the series.

$$n=\frac{z-a}{d}+1 \qquad =\frac{125-5}{5}+1$$

$$=24+1 \qquad =25$$

Or,

$$n=\frac{2S}{a+z} \qquad =\frac{2\times1625}{5+125}$$

$$=\frac{3250}{130} \qquad =25$$

Example.—There are 25 terms in a series, the first term of which is 5, the last 125, the common difference being 5. Find the sum of a series.

$$S=\frac{n(a+z)}{2} \qquad =\frac{25(5+125)}{2}$$

$$=\frac{25\times 130}{2} \qquad =\frac{3250}{2} \qquad =1625$$

Or,

$$S=n[a+\frac{d}{2}(n-1)] \qquad =25[(5+\frac{5}{2}(25-1)]$$

$$=25(5+60) \qquad =1625$$

Or,

$$S=n[z-\frac{d}{2}(n-1)] \qquad =25[125-\frac{5}{2}(25-1)]$$

$$=25[125-60] \qquad =25\times 65 \qquad =1625$$

Example.—Find the last term of a series : the first is 5, the common difference 5, and number of terms 25.

$$z=a+d(n-1) \qquad =5+5(25-1)$$

$$=5+120 \qquad =125$$

Or,

$$z=\frac{S}{n}+\frac{d}{2}(n-1) \qquad =\frac{1625}{25}+\frac{5}{2}(25-1)$$

$$=65+\frac{5}{2}(24) \qquad =65+60 \qquad =125$$

If the sum of the series is known, we could use another formula :

$$z=\frac{2S}{n}-a \qquad =\frac{2\times 1625}{25}-5$$

$$=\frac{3250}{25}-5 \qquad =130-5 \qquad =125$$

Example.—Find the first term of a series of 25 terms whose sum equals 1625, and last term is 125.

$$a=\frac{2S}{n}-z \qquad =\frac{2\times 1625}{25}-125$$

$$=130-125 \qquad =5$$

Or,

$$a=z-d(n-1) \qquad =125-5(25-1)$$
$$=125-120 \qquad =5$$

Or,

$$a=\frac{S}{n}-\frac{d}{2}(n-1) \qquad =\frac{1625}{25}-\frac{5}{2}(25-1)$$
$$=65-60 \qquad =5$$

Observe that the number outside the brackets is multiplied by the number within it. We do not subtract or add it (according to the sign which prefixes it) to the number to the left of it, for this would give an erroneous answer. In the last example, for instance:

$\frac{1625}{25}-\frac{5}{2}=65-2\frac{1}{2}=62\frac{1}{2}$, and $62\frac{1}{2}$ multiplied by $25-1=24$ would be 1500, obviously wrong.

Example.—Find the constant difference in a series of 25 terms, whose first term is 5, last term 125.

$$d=\frac{z-a}{n-1} \qquad =\frac{125-5}{25-1} \qquad =\frac{120}{24}=5$$

If the sum is also known:

$$d=\frac{(z+a)\,(z-a)}{2S-a-z} \qquad =\frac{(125+5)\,(125-5)}{2\times1625-5-125}$$
$$=\frac{130\times120}{3250-130} \qquad =\frac{15600}{3120}=5$$

Or,

$$d=\frac{2(S-an)}{n(n-1)} \qquad =2\left[\frac{1625-(5\times25)}{25(25-1)}\right]$$
$$=\frac{2(1625-125)}{25\times24} \qquad =\frac{3000}{600}=5$$

Geometrical Progression.—A series of numbers which increase or decrease by a *constant factor or common ratio* is known as a geometrical progression. For example, the series 3, 9, 27, 81, 243, 729, . . . is in geometrical progression, the constant factor being 3. A series such as $-\frac{1}{2}, \frac{1}{16}, -\frac{1}{128}, \frac{1}{1024}$ is also in geometrical progression, the difference in this case being $-\frac{1}{8}$, and the terms are alternately positive and negative.

Let a=the first term of the series, z the last term, n the number of terms, r the constant factor, and S the sum of the terms.

$$a=\frac{z}{r^{n-1}}.\quad a=S-r\,(S-z).\quad a=S\,\frac{r-1}{r^{n}-1}.\quad z=ar^{n-1}$$

$$z=S-\frac{S-a}{r}.\quad z=S\left(\frac{r-1}{r^{n}-1}\right)r^{n-1}.\quad r=\sqrt[n-1]{\frac{z}{a}}.\quad r=\frac{S-a}{S-z}$$

$$ar^{n}+S-rS-a=O.\quad S=a\,\frac{(r^{n}-1)}{r-1}.\quad S=a\,\frac{(1-r^{n})}{1-r}.\quad S=\frac{rz-a}{r-1}$$

$$S=\frac{z(r^{n-1})}{(r-1)r^{n-1}}.\quad n=\tfrac{1}{4}+\frac{\log z-\log a}{\log r}$$

$$n=1+\frac{\log z-\log a}{\log (S+a)-\log (S-z)}$$

$$n=\frac{\log\,[a+S(r-1)]-\log a}{\log r}$$

$$n=1+\frac{\log z-\log[zr-S(r-1)]}{\log r}$$

$$S=\frac{z^{n-1}\sqrt{z}-a^{\;-1}\sqrt{a}}{{}^{n-1}\sqrt{z}-{}^{n-1}\sqrt{a}}$$

The Geometric Mean of two quantities, A and $B=\sqrt{AB}$.

The reader will be able to understand the application of the formulæ for geometrical progression from the examples given for arithmetical progression.

Harmonical Progression.—Quantities are in harmonical progression when any three consecutive terms being taken, the first is to the third as the difference between the first and second is to the difference between the second and third. Thus, if a, b, c are the consecutive terms in a series, then if $a : c :: a\text{-}b : b\text{-}c$, then a, b, c are in harmonical progression, and if quantities are in harmonical progression their reciprocals are in arithmetical progression. The harmonic mean of two quantities A and B is $\dfrac{2AB}{A+B}$.

There is no simple formula for the sum of a harmonical

progression, but taking the ordinary harmonic series $1+\frac{1}{2}+\frac{1}{3}+\frac{1}{4}+\ldots\ldots\frac{1}{n}$, and calling its sum to n terms Sn, it can readily be shown approximately that $Log(n+1)<Sn<1+Log\ n$. In other words, Sn differs by less than 1 from $log\ n$. More exactly it can be shown, though proof is long and difficult, that where n is very large $Sn>log\ n$, by an amount known as 8.

Summary.—Arithmetical progression is a series of numbers which increase or decrease by a constant or common difference. Geometrical progression is a series of numbers which increase or decrease by a constant factor or common ratio. In a later chapter on Interest, the application of geometrical progression is given. The formulæ can also be applied, and probably find their most frequent application, in calculating speeds in machine-tool drives which include stepped pulleys.

Harmonical Progression is a series in which, when any three consecutive terms are taken, the first is to the third, as the difference between the first and second is to the difference between the second and third.

When the ratio of a geometrical progression is in excess of unity, it is known as an *increasing progression.* When it is less than unity, it is known as a *decreasing progression.*

EULER'S NUMBERS E

$$2/(e^x+e^{-z})=1+\Sigma(-1)^n E_n x^{2n}/(2n)\,!$$

$E_1=1$	$E_6=2{\cdot}702765\times10^6$
$E_2=5$	$E_7=1{\cdot}99360981\times10^8$
$E_3=6{\cdot}1\times10$	$E_8=1{\cdot}9391512145\times10^{10}$
$E_4=1{\cdot}385\times10^3$	$E_9=2{\cdot}404879675441\times10^{12}$
$E_5=5{\cdot}0521\times10^4$	

EULER'S CONSTANT

$$y=\underset{n=\infty}{\mathrm{Lt}}\left(\frac{1}{1}+\frac{1}{2}+\frac{1}{3}+\ \ldots\ +\frac{1}{n}-\log n\right)$$

$$={\cdot}5721$$

CHAPTER XI

Averages—Ratio and Proportion—Percentages

Averages.—*To find the average of a series of quantities the total of the similar quantities is divided by the number of quantities.* For example, the average earnings of ten men earning 30s., £2, 50s., £3, 70s., £4, 90s., £5, 110s., £6 respectively is £3, 15s.

If a train completes a journey of 90 miles in 3 hours, its average speed is 30 miles an hour. It will be seen that the sum of the terms is equal to the average multiplied by the number of terms.

Another example : What is the average of ·36, ·75, ·42, ·68, ·101, 3·75 ?

The sum of these items is 6·061, and there are 6 terms. Dividing 6·061 by 6 we obtain 1·01016.

Ratio and Proportion.—We often convey an idea of the size of an object or the length of time which is taken to perform a certain operation by comparing it with some other known object or period. For example, we could say that one car is three times the size of another, or that it takes twice as long to go from London to Birmingham as it does from London to Gloucester. This is the principle of proportion, and the relation which one quantity bears to another is called a ratio. When comparing two quantities of the same kind we consider how many times one quantity is contained in the other. We divide the first quantity by the second and the quotient expresses the ratio. For example, 2s. is to £1 as 1 is to 10. We have divided 2 into 20 and obtained 10 as the quotient. This ratio can be expressed as $\frac{20}{2}$, $20 \div 2$, or $20 : 2$. Dissimilar things cannot be expressed as a ratio.

When we desire to divide some number in a given ratio we add together the two terms of the ratio and express it as a denominator of the number. For example, divide 125 in the ratio 2 : 3. We add 3 and 2 together and obtain 5 ; thus $\frac{125}{5}=25$. We now multiply each term of the ratio by 25, thus obtaining 50 and 75, which represent the number 125 divided in the ratio 2 : 3.

Similarly, if 125 were to be divided in the ratio $\frac{1}{2}:\frac{1}{3}$ we reduce these fractions to a common denominator first. Thus, $\frac{1}{2}=\frac{3}{6}$, and $\frac{1}{3}=\frac{2}{6}$. Now divide 125 by the sum of the two numerators 3 and 2, or 5, and using 5 as a denominator and 3 and 2 as numerators, we split up 125 in the ratio $\frac{1}{2}:\frac{1}{3}$. Thus $\frac{3}{5}\times\frac{125}{1}=75$, and $\frac{2}{5}\times\frac{125}{1}=50$. Hence, as we shall see later, $\frac{1}{2}:\frac{1}{3}$ as 75 : 50. Note that we do not take $\frac{1}{2}$ of 125 and $\frac{1}{3}$ of 125 to express the ratio.

The ratio 3 : 6 is the same as 6 : 12, and we can express this in the usual way as 3 : 6 :: 6 : 12. In this example the ratios are equal, and the terms are hence said to be *in proportion*. Expressed as a rule, *four quantities are proportional when the ratio of the first to the second is equal to the ratio of the third to the fourth.* Thus, 3 : 4 :: 9 : 12, or 7 : 10 :: 35 : 50 are in proportion. The first and last terms of a proportional expression are called the *extremes*, and the two middle terms are called the *means*. It is important to note that the product of the extremes is equal to the product of the means, and this is a test of whether the terms are in proportion.

It will be observed from this that if three terms of a proportion are known the remaining one can be calculated. For example, find the second term of a proportion, the first, third, and fourth terms of which are 3, 6, and 12. Multiply 3 by 12 to obtain 36. Divide by the remaining term (6), and obtain the second term, 6.

Quantities are in *direct proportion* when the first is to the second as the second is to the third. Thus 2 : 4 : 8 is a direct proportion.

When the second term is equal to the third, each is a *mean proportional* to the other two. In the above example 6 is a mean proportional to 3 and 12.

When the ratio of the first to the second is the same as the second to the third, the latter is a third proportional to the other two.

The *geometrical mean* of two numbers is found by extracting the square root of their product, thus the geometrical mean of 4 and 16 is $\sqrt{4\times16}=8$.

The *arithmetical mean* is half the sum of two numbers. In the previous example the arithmetical mean would be $\frac{4+16}{2}=10$.

Percentages.—A percentage is a number or fraction with a denominator of 100, and it represents the rate of increase or reduction of one quantity with another quantity of the same kind. Thus, $10\%=\frac{10}{100}$ or $\frac{1}{10}$, and $17\frac{1}{2}\%=\frac{35}{200}$, or $\frac{7}{40}$. For example, suppose the seating capacity of a theatre is 500 and 300 people turn up, we say that $\frac{300}{500}\times\frac{100}{1}=60\%$ of the seating capacity is occupied. If we know that the seating capacity is 500, and we are told that the audience were 60 per cent. of the total capacity of the theatre, we use 100 as a denominator : thus, $\frac{60}{100}\times500=300$.

Another example. What is $33\frac{1}{3}$ per cent. of £24, 6s. 9d.? $\frac{33\frac{1}{3}}{100}=\frac{100}{300}=\frac{1}{3}$, and $\frac{1}{3}$ of £24, 6s. 9d. is £8, 2s. 3d.

CHAPTER XII

INTEREST—DISCOUNT—PRESENT VALUE—ANNUITIES

Simple Interest.—Money which is paid for the use of money loaned for a certain time, or invested at a fixed rate per cent., is known as *Interest*.

Let P=the *Principal* or money lent (expressed in pounds).
R=*Rate per cent.*, or the interest on £100 for one year.
A=the *Amount*, or principal+interest.
I=*Interest*.
T=*Time* in years.

Then:

$$A=P+I=\frac{P(100+RT)}{100}$$

$$I=\frac{PTR}{100}$$

$$P=\frac{100\,I}{RT}$$

$$R=\frac{100\,I}{PT}$$

$$T=\frac{100\,I}{PR}$$

$$P=\frac{100\,A}{100+RT}$$

It should be remembered that if the time is given in days, weeks, or months, the 100 in these equations should be multiplied by 365, 52, or 12 respectively.

Compound Interest.—The interest due at the end of each year is added to the principal in compound interest, and thus the principal increases from year to year as the interest is added.

Let A=*Amount*.
n=*Periods*, or number of years.
P=*Principal*.
r=*Interest* on £1 per year, or stated period.

Then :

$$n=\frac{\text{Log } A-\text{Log } P}{\text{Log } (1+r)}$$

$$A=P(1+r)^n$$

$$P=\frac{A}{(1+r)^n}$$

$$r=\sqrt[n]{\frac{A}{P}}-1$$

Discount and Present Value.—Discount represents the difference between the *Amount* and its *present value.* Expressed in another way, it is the interest on the present value for a given period.

Banker's discount, or *commercial discount*, is the interest charged on the amount of the debt, and it is hence greater than the true discount.

The present value of a sum of money which is due at some future date is that amount of money which, plus its interest from the present moment to the time when the payment is due, will equal the given amount.

The present value of £500, for example, due 12 months hence at 5 per cent. is £475, because the interest amounts to £25.

Let P = the present value.
$D1$ = banker's discount.
A = amount of debt.
r = interest on £1 for one year.
n = number of years.

$$A=P(1+nr)$$

$$A=\frac{D(1+nr)}{nr}$$

$$D=A-P=A-\frac{A}{1+nr}=\frac{Anr}{1+nr}$$

$$P=\frac{A}{1+nr}$$

$$D1=Anr$$

$$n=\frac{A-P}{Pr}$$

Present Value and Discount.—The present value or present worth of a sum of money due at some future date is equal to the sum of money which, plus its interest from the present time to the time when it becomes due for payment, will be equal to the given amount. Thus, the present value of £412 in nine months' time at 4 per cent. simple interest is £400.

True discount is the difference between the amount and the present value—the interest on the present value for a given period.

Banker's discount is interest on the amount of the debt, and thus is always greater than the true discounts.

In the example given the banker's discount would be £12, 7s. $2\frac{2}{5}$d., whereas the true discount is £12.

A=amount of debt.
D=true discount.
D_1=banker's discount.
P=present value.
r=interest on £1 for one year.
n=number of years.

$$A=\frac{D(1+nr)}{nr}$$

$$A=P(1+nr)$$

$$D_1=Anr.$$

$$r=\frac{A-P}{Pn}$$

$$n=\frac{A-P}{Pr}$$

$$P=\frac{A}{1+nr}$$

$$D=A-P=A-\frac{A}{1+nr}=\frac{Anr}{1+nr}$$

Annuities.—A fixed sum of money paid periodically is an annuity. The present value of an annuity which is to continue for n years at compound interest is :

$$P=\frac{1-(1+r)^{-n}}{r}N$$

In the case of a perpetual annuity the present value is found by making n infinite in this formula.

Hence in this case :

$$P=\frac{n}{R}$$

If an annuity is left unpaid for n years the whole amount A at simple interest (r=interest of £1 for one year) is :

$$A=nN+\frac{n(n-1)}{2}rn$$

At compound interest :

$$A=\frac{(1+r)^{n-1}}{r}N$$

It is not possible to calculate with any degree of reliability the present value of an annuity at simple interest.

Let mN=present value of an annuity.
N=annuity.

Then the annuity is worth m years' purchase.

The number of years' purchase (m) of a perpetual annuity is :

$$m=\frac{1}{r}=\frac{100}{\text{rate per cent.}}$$

In other words, an annuity is worth m years' purchase.

A debt which bears compound interest is reduced by an annual payment N.

The formula for finding the debt d after n years, and the time m when it will be fully paid is :

$$d=\frac{(Dr-N)\,(1+r)^n+N}{r}$$

$$m=\frac{Log\ N-Log\ (N-Dr)}{Log\ (1+r)}$$

When $n=Dr$ the debt will never be paid.

To find the annual payment N which will clear a debt D at r per cent. compound interest in n years :

$$N=\frac{Dr\,(1+r)^n}{(1+r)^{n-1}}$$

CHAPTER XIII

ALGEBRA

WHEREAS in ordinary arithmetical calculation we make use of the digits 1 to 0 to express quantities, in algebra we make use of letters of the alphabet as well as the digits. Figures can refer to yards, or feet, or any of the other units, such as 9 ft., 10 gals., 4 amperes, 3 acres, 2 lbs., 16s., £27, and so on. The signs which we use in connection with arithmetic are also used in algebra, with modifications and additions.

It is customary in algebraic expressions to use the letters a, b, c, etc., for known quantities, and x, y, z for unknown quantities. One of the difficulties which beginners in algebra always encounter is the correct use of the plus and the minus sign. For example, if we are considering the movement of a train, say, in one direction, we regard that as positive, whilst when the train travels in the reverse direction we regard that as negative. We should express the distance covered in the first direction or the speed at which the distance is covered, if those factors were known, as $+a$, and in the reverse direction as $-a$. In this connection the signs are used somewhat differently from their arithmetical use, for in the latter case the plus sign is merely used in addition sums, and the minus sign in subtraction sums. Moreover, it is necessary in algebra to get clearly established in the mind the fact that a quantity may be subtracted from another quantity smaller than itself. We must get accustomed to the fact that there are quantities less than nought. A man may have no money and we can say that his wealth is represented by o, but if he owes, say, £30, his wealth is represented by -30, and in temperatures we refer to degrees below freezing-point (Fahrenheit scale) by prefixing the $-$ sign to the number of degrees.

In algebra it is customary (although there are exceptions) to put the plus quantities first, and so eliminate the use of the plus sign. Thus, $x+y$ means $+x+y$, and $x-y$ means $+x-y$.

It is convenient to regard the subtraction of one quantity from another as the addition of a negative quantity to a positive quantity. For example, by adding $+15$ and -24 we should obtain as an answer -9. This is known as the *algebraic sum.* It is important to get this point fixed in the mind. Whereas in arithmetic the minus sign means subtract, in algebra it means the addition of a negative quantity.

Also, in arithmetic the multiplication sign is used to indicate the two quantities that are to be multiplied together. Sometimes in algebra we also use the multiplication sign and in others it is omitted. For example, ab means that a and b are multiplied together, and the expression is more convenient than $a \times b$. If ordinary digits are included in an expression such as $25 \times a \times b \times c$ we should express the quantity as $25abc$. Algebra thus differs from arithmetic, for 9×12 could not be expressed as 912.

When two quantities are separated by a plus sign and they have to be multiplied by two other quantities separated by a plus sign the expression is written : $(x+y) \times (a+b)$, or more simply $(a+b)\ (x+y)$.

Quantities such as $x+y-z$, or $17aby$, are known as *quantities*, or *algebraic expressions.*

A quantity such as $7x$ means that x is to be multiplied by 7. The numerical part of the quantity is called the *coefficient.* The quantity itself, when it consists of a coefficient and a letter is called a *term.*

Powers of quantities are expressed as in arithmetic ; thus x^3 means the third power of x, x^2 means the square of x, $\sqrt{x}$ means the square root of x, and so on. As in logarithms the number indicating the power is called the *index* or *exponent.* In order that the reader may understand signs used in algebra the following examples are given :

$$x^2+y+z$$

If $x=3$, $y=7$, and $z=6$, the value of the expression is

$$\begin{aligned} x^2 &= 9 \\ y &= 7 \\ z &= 6 \\ \hline & \quad 22 \end{aligned}$$

Similarly, x^2-y+2z, and giving the same values, would provide $9-7+12$. Adding the positive quantities produces 21, and subtracting the negative quantity gives the answer 14.

Find the value of $\frac{3x\times y+z}{ab-c^2}$, where $x=2$, $y=3$, $z=6$, $a=7$, $b=5$, and $c=4$.

We have $\frac{3\times 2\times 3+6}{7\times 5-16}=\frac{24}{19}$

Addition.—When adding algebraic quantities all the like quantities are added together, particular note being taken of their signs. When the quantities have a similar sign, the coefficients are added and the letters annexed. For example : $5x+12x=17x$. Again, $3x+12x-3x+4a+4b=12x+4a+4b$.

Subtraction.—Here the terms are arranged as in addition, but the signs of the terms to be subtracted are changed. Thus, the minus signs in the terms to be subtracted are changed to plus and the plus terms are changed to minus. Hence, in subtracting $7x$ from $10x$, we write : $10x-7x=3x$.

Again, from $3x+2y-3z$ subtract $2x-z$.

$$\begin{array}{l} 3x+2y-3z \\ 2x \qquad\ \ -z \\ \hline x+2y-2z \\ \hline \end{array}$$

It will be observed that the process of subtracting a negative quantity really means the adding of an equivalent positive quantity.

Multiplication.—We have already seen that the multiplication sign can often be omitted in algebra, and that xy really means $x\times y$. Also, $x(x-y)$ is the same as $x\times(x-y)$. In multiplication like signs produce a positive result. When unlike signs are multiplied together the answer is negative. For example : $5x\times 3x=15x^2$, and $5x\times(-3x)=-15x^2$.

In multiplying terms which are powers, such as $2x^2\times 3x^3$, the coefficients are multiplied together, but the indices are added; thus, in this example, the answer would be $6x^5$.

A *continued product* is obtained when a number of quantities are multiplied together. For example : The continued product of $2x$, $3y$, and $4z$ is $24xyz$.

Division.—The same rules as in arithmetic apply. For example: Divide $27xy^2$ by $3xy$. This should be expressed:

$$\frac{27xy^2}{3xy}=9y$$

The result of any division sum can be checked by multiplying the divisor and quotient together and this should equal the dividend. The dividend and divisor are arranged according to the powers of the symbol, and in descending order. Thus:

$$+2xy+y^2)x^5+5x^4y+10x^3y^2+10x^2y^3+5xy^4+y^5(x^3+3x^2y+3xy^2+y^3$$
$$x^5+2x^4y+x^3y^2$$
$$3x^4y+9x^3y^2+10x^2y^3$$
$$3x^4y+6x^3y^2+3x^2y^3$$
$$3x^3y^2+7x^2y^3+5xy^4$$
$$3x^3y^2+6x^2y^3+3xy^4$$
$$x^2y^3+2xy^4+y^5$$
$$x^2y^3+2xy^4+y^5$$

Here it will be seen that x^2 has been divided into the first term x^5 to produce x^3. Then the whole of the divisor is multiplied by x^3 to produce $x^3+2x^4y+x^3y^2$, whilst x^3 itself is placed in the quotient; and so on.

Here are further examples:

$$a+b)a^3+3a^2b+3ab^2+b^3(a^2+2ab+b^2$$
$$a^3+a^2b$$
$$2a^2b+3ab^2$$
$$2a^2b+2ab^2$$
$$ab^2+b^3$$
$$ab^2+b^3$$

Or,

$$a^2+2ab+b^2)a^3+3a^2b+3ab^2+b^3(a+b$$
$$a^3+2a^2b+ab^2$$
$$a^2b+2ab^2+b^3$$
$$a^2b+2ab^2+b^3$$

$$
\begin{array}{l}
a+b)ax+ay-az+bx+by-bz(x+y-z \\
\quad ax \qquad\quad +bx \\
\hline
\quad ay-az+by \\
\quad ay \qquad +by \\
\hline
\quad -az-bz \\
\quad -az-bz \\
\hline
\end{array}
$$

Brackets.—Parts of an algebraic expression are often grouped by means of brackets ( ), and where brackets are used within a bracket, square brackets [ ] are used on the outside and the normal bracket within. Sometimes it is necessary also to use the brace $\{\ \}$. In other instances a line is placed over quantities, and this line has the same value as a bracket. Hence, $\overline{x+y}$ is the same as $(x+y)$, and $\overline{x+y}\times\overline{x-y}$ is the same as $(x+y)(x-y)$.

Example: $5x-(2a-3b)$. The expression means that we must subtract $2a-3b$ from $5x$.

When a minus sign precedes the bracket, it changes the sign of all terms within the bracket. The minus sign is equivalent to -1, and every term within the bracket is thus multiplied by -1. So, in the example given, $2a$ becomes $-2a$, and $-3b$ becomes $+3b$, and the complete expression now reads $5x-2a+3b$.

The positive sign outside a bracket does not affect the signs within.

Further examples:

$$5x+(3a-4b+5c)=5x+3a-4b+5c$$
$$5x-(3a-4b+5c)=5x-3a+4b-5c$$

When multiple brackets are used, each form of bracket indicates that all terms between them must be considered as one quantity and therefore must be multiplied, divided, subtracted or added as a whole according to the signs which precede the terms.

Thus, to multiply $x+y$ by $a+b$ we write:

$$(x+y)(a+b)$$

If we wish to subtract a quantity from this, as $2d$, we write:

$$(x+y)(a+b)-2d$$

and if this quantity is to be multiplied by a further term, say $5x$, we write :

$$5x\left\{(x+y)(a+b)-2d\right\}$$

Further multiplication, addition, subtraction or multiplication brings in the use of square brackets. Thus :

$$2[5x\left\{(x+y)(a+b)-2a\right\}-n]$$

It is usual, in evaluating an expression of this sort, to remove the inner brackets first ; thus :

$$2[5x\left\{ax+ay+bx+by-2a\right\}-n]$$

Here $x+y$ has been multiplied by $a+b$ to produce $ax+ay+bx+by$.

Next the braces ; multiply $ax+ay+bx+by-2a$ by $5x$, producing :

$$5ax^2+5axy+5bx^2+5bxy-10ax$$

We now have

$$2(5ax^2+5axy+5bx^2+5bxy-10ax-n)$$

and multiplying that by 2 produces

$$10ax^2+10axy+10bx^2+10bxy-20ax-2n$$

Further Examples.—Add : $2a+4b+3c+8d$, $a+b-c-d$, $18a-3b+3c-4d$, and $6b+5c+4d$.

$$\begin{array}{r} 2a+4b+\ 3c+8d \\ a+\ b-\ \ c-\ d \\ 18a-3b+\ 3c-4d \\ 6b+\ 5c+4d \\ \hline \text{Answer : }\ 21a+8b+10c+7d \\ \hline \end{array}$$

Add : $a^2+3b-c+5d$, $3b+c$, a^2+3c, $-a^2+c-d$, $c-d$, and $4a^2-2b+c-3d$.

$$\begin{array}{rrrr} a^2 & +3b & -\ c & +5d \\ & 3b & +\ c & \\ a^2 & & +3c & \\ -a^2 & & +\ c & -\ d \\ & & c & -\ d \\ 4a^2 & -2b & +\ c & -3d \\ \hline \end{array}$$

Answer : $5a^2+4b+6c$

Add: $2ab-3ac+4bx-3ay+3$, $ac+2ay-4$, $-ab-2ac-bx-ay-4$, $5ac+2$, and $ab+ay$.

$$
\begin{array}{rrrrr}
2ab & -3ac & +4bx & -3ay & +3 \\
 & ac & & +2ay & -4 \\
-ab & -2ac & -bx & -ay & -4 \\
 & 5ac & & & +2 \\
ab & & & +ay & \\
\hline
2ab & +ac & +3bx & -ay & -3
\end{array}
$$

Answer: $2ab+ac+3bx-ay-3$

If in the last example $a=2$, $b=3$, $c=4$, $x=5$, and $y=6$, the sum would become:

$$
\begin{array}{rr}
12-24+60-36+3=75-60= & 15 \\
8+24-4=32-4= & 28 \\
-6-16-15-12-4 = & -53 \\
40+2 = & 42 \\
6+12 = & 18 \\
\hline
 & 50
\end{array}
$$

or, by substituting in the answer:

$$12+8+45-12-3=65-15=50.$$

Subtract: $3x^2-2xy+9$ from $5x^2-xy+ab-2$.

$$
\begin{array}{rrrr}
5x^2 & -xy & +ab & -2 \\
3x^2 & -2xy & & +9 \\
\hline
2x^2 & +xy & +ab & -11
\end{array}
$$

Answer: $2x^2+xy+ab-11$

Subtract: $9a^2+4ab+b^2$ from $a^2+2ab+b^2$.

$$
\begin{array}{rrr}
a^2 & +2ab & +b^2 \\
9a^2 & +4ab & +b^2 \\
\hline
-8a^2 & -2ab &
\end{array}
$$

Answer: $-8a^2-2ab$

Divide: $-x+y$ by -1.

Answer: $x-y$.

Express $-x+y$ in another form.

Answer: $-(x-y)$.

The reader should practise multiplication in algebra and check the result by dividing the product by one of the two quantities which, multiplied together, produced the products. The answer should be the other quantity.

Thus : Multiply $ax+3by$ by $4a+3b$.

$$\begin{array}{l} ax+3by \\ 4a+3b \\ \hline 4a^2x+12aby \\ \qquad\qquad +3abx+9b^2y \\ \hline \end{array}$$

Answer : $\underline{4a^2x+12aby+3abx+9b^2y}$

We can now check this back :

$$ax+3by)4a^2x+12aby+3abx+9b^2y(4a+3b$$
$$\begin{array}{l} 4a^2x+12aby \\ \hline 3abx+9b^2y \\ 3abx+9b^2y \\ \hline \end{array}$$

Or, $$4a+3b)4a^2x+12aby+3abx+9b^2y(ax+3by$$
$$\begin{array}{l} 4a^2x \qquad +3abx \\ \hline 12aby+9b^2y \\ 12aby+9b^2y \\ \hline \end{array}$$

CHAPTER XIV

Simple Equations

Algebra is of the greatest use in expressing a problem in its simplest terms ; it is a system of arithmetical shorthand, more graphic than words and enabling the relation of the quantities to one another to be seen at a glance. The problem is converted into *symbols*, or *symbolical expressions.*

Thus, the output of a machine may be stated as x, and of five machines of similar type as $5x$; or if the rate of output of a machine is increased by five times, its output would be $5x$.

Two numbers may differ by, say, 17. Let x denote the larger number and y the smaller.

Hence $x-y=17$.

A car travelling a distance of x miles at a speed of y miles an hour would take $\frac{x}{y}$ hours.

Now, in algebraic calculations certain symbols are assigned to certain quantities. Thus t=time, V and v=velocity, s= distance, or space, or sum ; g=acceleration due to gravity= 32·2 ft. per sec./per sec. These letters convey what they stand for at a glance, because they are the initial letters of the quantities.

Any arithmetical or algebraic expression such as $x+y=9$, $3\times 8=24$ is called an *equation*, since the terms each side of the = sign are *equal*, or *equated.* All expressions which are separated by the equality sign are therefore equations, whether the expressions are arithmetical or algebraic.

By means of algebraic equations we are able to find the value of the unknown quantities, and this process is known as *solving the equation*, the answer being the *root* or *solution.*

If the problem does not involve the power of any unknown quantity, it is termed a *simple equation* ; if one or more of the unknown quantities are squared, the problem is a *quadratic equation* ; if one or more unknown quantities are cubed, the problem is a *cubic equation.* If the fourth power is involved, it is a *quartic.*

When the equation relates to an algebraic operation only the expression is known as an *identity*. Thus $(a+b)^2=a^2+2ab+b^2$ is an identity, for we have evaluated the expression $(a+b)^2$, and reached a result beyond which it is impossible to proceed.

Continuing with simple equations, let us take the simple algebraic expression $5a+9=19$. In this expression it is clear that 5 times the unknown number plus $9=19$. It is apparent that a equals 2, and this simple example is sufficient to illustrate the process of solving an equation.

Now it is important to remember that the value of an equation is not altered by multiplying, dividing, subtracting, or adding to each side of the equation. Whatever is done to one side of the equation must be done to the other. We can, if it is convenient to solve the equation in that way, raise each side of the equation to the second or third power, or to the nth power, which is another way of saying that if you multiply, divide, subtract from, add to, or raise to a power the left-hand side of the equation, we must multiply, divide, subtract from, add to, or raise to the same power the right-hand side.

Although mentally we have found that the value of a in the above simple equation is 2, it would, in fact, be solved in the following way:

$$5a+9=19$$

Now subtract 9 from both sides, thus:

$$(5a+9)-9=19--9$$

Subtracting, $5a=10$.

Dividing throughout by the coefficient 5:

$$a=2$$

The balance of the equation is not destroyed, even if all the signs on each side are changed, thus:

$$-5a-9=-19$$

This merely means that we have multiplied the original equation by -1.

Another method of solving the equation would be to transpose the figure 9 from one side of the equation to the other, thus:

$$5a+9=19$$
$$5a=19-9$$

This boils down to the well-known algebraic rule :

All the unknown quantities are transposed to one side and all the known quantities to the other, dividing, if necessary, by the coefficient of the unknown quantity.

Here are some examples :

$$3x+5=20$$
$$3x=20-5$$
$$3x=15$$
$$x=5$$
$$2(x+4)=24$$
$$2x+8=24$$
$$2x=24-8$$
$$2x=16$$
$$x=8$$
$$3x^2-7=41$$
$$3x^2=48$$
$$x^2=16$$
$$x=\sqrt{16}$$
$$=4$$
$$(a+b)(a+b)=25$$
$$a^2+2ab+b^2=25$$

This is an identity, and numerical values for a and b cannot be found (except by trial and error) unless we know the value of one of the unknowns. We can, of course, observe that, in this case, $a+b=5$.

$$5x-7=2x+2$$
$$5x-2x=2+7$$
$$3x=9$$
$$x=3$$

Note that, when transposing from one side to another, the signs are changed. Negative signs become positive and positive signs become negative.

Thus in the last example $+2x$ becomes $-2x$, and -7 becomes $+7$. This rule is most important.

$$3a^2-10a=8a+a^2$$
$$3a^2-a^2=10a+8a$$
$$2a^2=18a$$
$$a^2=9a$$

Dividing by a $\quad a=9$

$$\frac{x}{3}+\frac{x}{9}=4$$

Get rid of the fractions by multiplying both sides of the equation by 9.

$$9\left(\frac{x}{3}\right)+9\left(\frac{x}{9}\right)=9\times 4$$
$$3x+x=36$$
$$4x=36$$
$$x=9$$

$$\tfrac{3}{5}\left(x+10\right)-\tfrac{2}{3}\left(x-2\right)=\frac{5x+5}{6}+2$$

Multiply both sides by 30 to eliminate the fractions :

$$18(x+10)-20(x-2)=25x+25+60$$
$$18x+180-20x+40=25x+25+60$$
$$18x-45x=-180-40+25+60$$
$$-27x=-135$$

Divide throughout by -27 :

$$x=5$$

The reader will become facile at solving equations if he practises making up simple problems and expressing the conditions by symbols.

Here are some examples :

The difference between two numbers is 7. Their sum is 13. What are the numbers ?

Let x=the larger number, and y the smaller.

$$\text{Then } x+y=13$$
$$x-y=7$$

Subtracting :

$$2y=6$$
$$y=3$$

Substituting this value in either of the above equations :

$$x+3=13$$
$$x=13-3$$
$$=10$$
$$\text{or } x-3=7$$
$$x=7+3$$
$$=10$$

Problems.—Find three numbers, such that the largest is equal to the sum of the other two ; the ratio of the latter is as 2 to 3. If 10 be added to each number the largest number will be 5 more than half the sum of the other two numbers.

As the two smaller numbers are in the ratio of 2 to 3, let $3x$ and $2x$ represent them ; $5x$ will therefore represent the largest number.

Add 10 to each of the numbers :

$$5x+10$$
$$3x+10$$
$$2x+10$$

Now :

$$5x+10=\tfrac{1}{2}(2x+3x+20)+5$$
$$\therefore\ 5x+10=\tfrac{1}{2}(5x+20)+5$$
$$\text{Hence } \tfrac{5}{2}x=5$$
$$x=2$$

The numbers are therefore 10, 6, and 4.

The result may be checked by substituting these values in the equation. Thus :

$$10+10=\tfrac{1}{2}(4+6+20)+5$$
$$20=\tfrac{1}{2}(30)+5$$
$$20=15+5$$

The ratio of two numbers is as 2 to 3, and their sum is 30. Find the numbers.

Let $2x$=one number.
and $3x$=the second number.

$$\text{Then } 3x+2x=30$$
$$5x=30$$
$$x=6$$

The numbers are therefore 12 and 18.

A stone dropped from a window takes 3 secs. to reach the ground. Find the height from which it was dropped. We must here remember that the velocity of a freely falling body is 32·2 ft. per sec./per sec. ; and the relation between distance fallen and time taken is $s=\frac{1}{2}gt^2$, where s=distance in ft., t= time in secs., and g=32·2.

So :

$$s=\tfrac{1}{2}\times 32{\cdot}2\times 3^2$$
$$=16{\cdot}1\times 9$$

Height of window=144·9 ft.

In what time will a stone drop a distance of 144·9 ft.?

$$144{\cdot}9=\tfrac{1}{2}\times 32{\cdot}2\times t^2$$
$$t^2=\frac{144{\cdot}9}{16{\cdot}1}$$
$$t=\sqrt{9}=3 \text{ secs.}$$

By adding 4 to a number, and dividing the result by 5, the result is one-fourth of the original number. What is the number ?

$$\frac{x+4}{5}=\frac{x}{4}$$

Multiply each side by 20 :

$$4x+16=5x$$
$$5x-4x=16$$
$$x=16$$

Pythagoras' rule for the length of the hypotenuse of a right-angled triangle is base^2+perpendicular^2=hypotenuse^2. The base of a right-angled triangle is 3 ins. and the hypotenuse 5 ins. What is its height ?

$$b^2+p^2=h^2$$
$$9+p^2=25$$
$$p^2=25-9$$
$$p=\sqrt{16}$$
$$=4$$

One number is one-fifth of another, and their sum is 30. What are the numbers ?

Let x=the smaller number.

Then $5x$=the larger number.

$$5x+x=30$$
$$6x=30$$
$$x=5$$

The two numbers are thus 5 and 25.

The foregoing are simple equations, involving the use of one unknown quantity. It is, of course, impossible to lay down specific rules for every type of problem. The reader must learn the various tricks of calculation to enable him to solve a problem. For example, we have seen that the square of two quantities separated by a plus sign is the square of each quantity, plus twice the two quantities multiplied together. Put in the form of an equation :

$$(x+y)^2=x^2+2xy+y^2$$

Here is an example of the use which may be made of this rule :

The product of two numbers is 20. The sum of their squares is 41. What are the numbers ?

$$xy=20$$
$$\therefore 2xy=40$$
$$x^2+y^2=41$$

By addition $x^2+2xy+y^2=81$

Factorising $(x+y)(x+y)$, or $(x+y)^2=81$.

Taking the square root of both sides, $x+y=9$.

But $xy=20$

Therefore $y=\frac{20}{x}$

Substituting for y:

$$x+\frac{20}{x}=9.$$

Multiplying both sides by x:

$$x^2+20=9x$$
$$x^2-9x+20=0$$

Factorising $(x-4)(x-5)=0$

Therefore, either $x-4=0$ or $x-5=0$

and $x=4$ or 5

Substituting in $x+y=9$

$y=$ either 5 or 4.

If we had also been given the difference of the two numbers (1), we should proceed thus:

$$xy=20$$
$$\therefore 2xy=40$$
$$x^2+y^2=41$$

Hence $x^2+2xy+y^2=81$

Factorising $(x+y)(x+y)=81$

Taking square root of both sides, $x+y=9$

But $x-y=1$

By addition $2x=10$

Substituting in $x+y=9$

$5+y=9$

$y=9-5=4$.

Here are some examples for practice. Evaluate y in the following equation:

$$\frac{11y-13}{25}+\frac{17y+4}{21}+\frac{19y+3}{7}=28\tfrac{1}{7}+\frac{5y-25\frac{1}{3}}{4}$$

Answer: 8.

Remember, it is always advantageous to clear the fractions.

Give an expression for x in the following equations :

$\frac{x-y}{x+y}+\frac{3a-x}{2a+x}=0$ Answer : $x=\frac{-ay}{5a-2y}$

$\frac{a-b}{x-c}=\frac{a+b}{x+2c}$ Answer : $\frac{3ac-bc}{2b}$

$\frac{ax}{b}+b=\frac{bx}{a}+a$ Answer : $\frac{ab}{a+b}$

$\frac{1}{x-a}+\frac{1}{x-b}=\frac{2}{x}$ Answer : $\frac{2ab}{a+b}$

$\frac{x}{ab}+\frac{x}{bc}+\frac{x}{ac}=a+b+c$ Answer : abc

Examples.—A man is older than his wife by 20 years, and 10 years ago the man was twice as old as his wife. What are their ages ?

Answer : 30 and 50.

At what time between eight and nine do the hands of a watch make an angle of 180° with each other ?

Answer : $10\frac{10}{11}$ mins. past 8.

What number is that which when multiplied by 3 will be in excess of 56 as much as it is now short of it ?

Answer : 28.

A dry cell has a voltage of 1·25 volts, and it is connected into a circuit whose resistance is 40 ohms. What is the current passing through the wire ? The internal resistance of the cell is 3 ohms. (*Note* : volts÷amps.=ohms.)

Answer : 0·02906 amperes.

Smith can complete a given piece of work in 9 hours. Jones takes twice that time, whilst Robinson does three-quarters as much work as Smith in 1 hour. How long will it take Smith, Jones, and Robinson, working together, to complete the piece of work ?

Answer : 4 hours.

The value of a fraction is $\frac{1}{2}$ when 4 is added to its numerator. The value of the fraction is $\frac{1}{4}$ if 6 is added to the denominator. What is the fraction ?

Answer : $\frac{7}{22}$.

If Smith gives 6s. to Robinson, the latter will have twice as much as Smith ; but if Robinson gives 3s. to Smith the latter will have the same amount as Robinson. What money has Smith and Robinson ?

Answer : 24s. and 30s.

CHAPTER XV

Simultaneous Equations

Pairs or sets of equations in which the same unknown symbols appear, which are assumed to possess the same values throughout, are *simultaneous equations*. By giving values to one of the unknown quantities values which correspond can be found for the other unknown quantity.

For example :

$$4x-3y=2$$

This really means that we must find two numbers of such a value that three times the second, subtracted from four times the first, equals 2. Now we have previously seen that we can *transpose* from one side to the other, and thus $4x-3y=2$ is equal to $4x=3y+2$.

Now give successive values of 1, 2, 3, and so on to y, so obtaining corresponding values of x. Thus :

$$y=1,\ 4x=5;\ \therefore\ x=\frac{5}{4}$$

$$y=2,\ 4x=8;\ \therefore\ x=\frac{8}{4}$$

$$y=3,\ 4x=11;\ \therefore\ x=\frac{11}{4}$$

$$y=4,\ 4x=14;\ \therefore\ x=\frac{14}{4}$$

Similarly, we can give values to x and find corresponding values of y. Later on we shall see the value of this when we come to plot graphs.

In the above equation we have found corresponding values, and so if we had a second equation, such as $6x-5y=2$, and give values to either x or y, corresponding values of the other unknown quantity will be obtained, and we can compile a similar table of values of x and y, as above. A comparison of the two sets of values will show that only one pair of values of x and y satisfies both equations.

Pairs of equations such as

$$4x-3y=2$$
$$2x+3y=28$$

in which the same values of the unknown apply are, as we have seen, *simultaneous equations.*

To find the value of two unknown quantities we have established that it is necessary to have two equations.

In the solution of simultaneous equations we must have as many independent equations as there are unknowns to be found. Therefore, if there are three unknowns we should require three distinct equations ; if there are four unknowns, as w, x, y, and z, we should require four equations.

Elimination.—When these equations are given it is possible by a process of *elimination* to obtain other equations, in which some of the unknowns do not occur. There are two methods generally adopted in solving simultaneous equations containing two unknown quantities. The first is *to find the value of one unknown in terms of the other unknown, and then to substitute the value so found in the other equations.*

The second is by *multiplication or division,* to make the coefficients of one of the unknowns the same in the two equations, when by subtraction or division one of the unknown quantities is eliminated, thus leaving only one unknown, the value of which will be found by the methods already described.

Example :

$$5x-2y=10$$
$$3x-\ \ y=\ 7$$

Applying the second method, multiply the first equation by 3, and the second by 5. This will produce :

$$\begin{array}{rr} & 15x-6y=30 \\ \text{Subtract} & \underline{15x-5y=35} \\ & -\ \ y=-5 \\ \text{Multiply throughout by } -1 & y=\ \ 5 \end{array}$$

Now substitute the value of 5 in the first equation, and we have

$$5x-10=10$$

from which $5x=20$ (transposing the -10 to the right-hand side of the equation to produce 20).

Hence, $x=4$.

By the first method $5x=10+2y$

$$\text{and} \quad x=\frac{10+2y}{5}$$

$$\text{From this } 3x=\frac{30+6y}{5}$$

Now substitute this value in the second equation $3x-y=7$.

$$\frac{30+6y}{5}-y=7$$

Next multiply both sides of the equation by 5 to get rid of the fraction.

$$30+6y-5y=35$$

$$\text{from which } 30+y=35$$

$$y=5$$

Substituting this value, the value of x can be found by the method already described.

It is very necessary to become adept at solving simultaneous equations. Here are some examples :

$$6x+15y=69$$

$$8x-6y=14$$

Answer : $x=4$, $y=3$

$$2x+\frac{2y-22}{3}=30$$

$$\frac{3x-15}{4}+6y=108$$

Answer : $x=13$, $y=17$

$$\frac{2}{x}+y=1$$

$$\frac{1}{x}+2y=1\tfrac{1}{4}$$

Answer : $x=4$, $y=\frac{1}{2}$

$$(a-b)x+(a+b)y=a^2-b^2$$

$$(a+b)x-(a-b)y=2ab$$

Answer : $x=\frac{1}{2}(a+b)$, $y=\frac{1}{2}(a-b)$

If 5 be added to the numerator of a fraction its value will be $\frac{1}{2}$, and if 1 be subtracted from the denominator, its value will be $\frac{1}{5}$. Find the fraction. The fraction is thus $\frac{x}{y}$. Add 5 to the numerator :

$$\frac{x+5}{y}=\frac{1}{2}$$

Now subtract 1 from the denominator,

$$\frac{x}{y-1}=\frac{1}{5}$$

Therefore $\dfrac{2x+10}{y}=1$

By cross-multiplication $2x+10=y$

Or $2x-y=-10$

From the second equation $5x=y-1$

By transposition we obtain $y-2x=10$

and $y-5x=1$

$$3x=9$$

$$x=3$$

Substituting this value in any one of the equations we find that $y=16$, and hence the fraction is $\frac{3}{16}$.

If 20 be added to a certain number and the result divided by 12, the result is one-eleventh of the original number. Find the number.

$$\frac{x+20}{12}=\frac{x}{11}$$

$$12x=11x+220$$

$$12x-11x=220$$

$$x=220$$

CHAPTER XVI

PERMUTATIONS AND COMBINATIONS

I NOW deal with the question of permutations and combinations, since, before the principles of the binomial theorem can be understood, it is necessary to understand this branch of calculation. A *permutation* is the number of different arrangements that can be made of a number of quantities.

If V represents the variations of n things taken r together :

$$V=n(n-1)(n-2)(n-3) \text{ to } r \text{ factors.}$$

Here is an example :

A licence is numbered BA3456 ; how many are there of a similar numbering system ? Counting the digits and letters we can regard the number as being composed of six compartments. The first two can be occupied by letters, and the remaining four by digits. It is obvious that the first compartment can be filled in 26 different ways, because there are 26 letters in the alphabet. If we put A in the first compartment there are 26 ways, similarly, of filling the second compartment, and similarly with B in the first compartment, and so on. Therefore, there are $26+26+26+26 \ldots$ or 26×26 different ways of filling the first two compartments. The digits are 1, 2, 3, 4, 5, 6, 7, 8, 9, 0, and therefore it is apparent that there are 10 different ways of filling each of the four remaining compartments, or 10^4 ways of filling the whole four compartments. Hence the licences, using four digits and two letters as in the example given, can be numbered in $26^2\times10^4$ different ways. The answer therefore is that in this series of licences there are 6,760,000 differently numbered licences.

Here is another example :

There are four signalling flags. In how many ways can these be arranged one above the other ?

Obviously any flag can be placed at the top, and when a particular flag has been so placed there are three ways of filling the space below. Therefore, the two upper positions for the flags can be filled in 4×3 different ways, and hence the whole series of flags can be arranged in $4 \times 3 \times 2 \times 1$ different ways, or 24 different ways in all. In other words, the arrangement is *factorial* 4, and is usually written mathematically as $\underline{|4}$. Sometimes the product of a series such as $4 \times 3 \times 2 \times 1$ is written 4 !, and the latter is frequently used.

Thus :

$$3\,! = 3 \times 2 \times 1 = 6$$
$$4\,! = 4 \times 3 \times 2 \times 1 = 24$$
$$5\,! = 5 \times 4 \times 3 \times 2 \times 1 = 120$$
$$6\,! = 6 \times 5 \times 4 \times 3 \times 2 \times 1 = 720$$

Now the factorial symbol ! or $\underline{|\ }$ can only be used in connection with a positive integer.

It is convenient to define 0 ! or $\underline{|0}$ as equal to unity.

Another example :

If there are six tin boxes, in how many different ways can any four of them be placed one above the other ?

From our previous reasoning it is obvious that the answer is $6 \times 5 \times 4 \times 3$. Thus the number of permutations of six things taken four at a time and arranged in a row, or one above the other, is 360.

Usually the permutations of n things taken r at a time is written ${}^nP_r$, which means the product of r factors decreasing by one at a time and beginning with n. The last factor will be $n-r+1$, and so we may rewrite the formula given above :

$${}^nP_r = n(n-1)(n-2)(n-3)\ .\ .\ .\ (n-r+1)$$

It is convenient in calculations involving factorial notation not to multiply large products, but to leave them in factorial form. Here are some examples :

The index number of a motor car is ABY430. How many such numbers are there containing three letters and one digit, three letters and three digits, and three letters and two digits ?

Answer : (1) $26^3 \times 9$; (2) $26^3 \times 9^3$; (3) $26^3 \times 9^2$.

A factory has n football teams. Each team plays every other team once ; how many matches do they play ?

Answer : $\frac{1}{2}n(n-1)$.

A bus service has 10 ticket stages. How many different stage tickets must be printed for this service ?

Answer : 90.

Combinations.—If there are seven equally good men out of which a team of three are to be provided, how many different teams could be selected ? It is apparent here that the arrangement of the men within the team does not matter. Therefore it becomes a problem involving the selection of different *groups* and not the formation of *different arrangements*. Each of these groups is hence termed a *combination* of n things taken r at a time. The number of such combinations is written ${}^nC_r$.

$$ {}^nC_r = \frac{{}^nP_r}{\underline{|r}} = \frac{n(n-1)(n-2)(n-3)\ .\ .\ .\ (n-r+1)}{\underline{|r}} $$

It is very necessary that the reader should understand the *difference between a combination and a permutation.*

Summarising, *a permutation is the order in which things are taken into account, whereas in a combination the order does not matter.* Hence, xyz, xzy, yxz, yzx, zxy, and zyx are different permutations of the same combination.

Example : How many different selections of four letters can be made from the letters u, v, w, x, y, z, without restriction, and secondly, if u must be in each selection ?

Also, if u and y must be in each selection.

Answer : (1) 15 ; (2) 10 ; (3) 6.

In a game of whist 52 cards are dealt in four hands of 13 cards. How many different arrangements of the cards are possible ?

Answer: $\dfrac{\underline{|52}}{(\underline{|13})^4}$

CHAPTER XVII

The Binomial Theorem

We have already seen that $(a+b)^2=a^2+2ab+b^2$, and that $(a+b)^3=a^3+3a^2b+3ab^2+b^3$.

Similarly $(a+b)^4=a^4+4a^3b+6a^2b^2+4ab^3+b^4$.

Proof: $(a+b)^2=(a+b)(a+b)$

$$\begin{array}{l} a+b \\ a+b \\ \hline a^2+ab \\ \quad +ab+b^2 \\ \hline a^2+2ab+b^2 \\ \hline \end{array}$$

$(a+b)^3=(a+b)(a+b)(a+b)$

$$\begin{array}{l} a^2+2ab+b^2 \\ a+b \\ \hline a^3+2a^2b+ab^2 \\ \qquad a^2b+2ab^2+b^3 \\ \hline a^3+3a^2b+3ab^2+b^3 \\ \hline \end{array}$$

$(a+b)^4=(a+b)(a+b)(a+b)(a+b)$

$$\begin{array}{l} a^3+3a^2b+3ab^2+b^3 \\ a+b \\ \hline a^4+3a^3b+3a^2b^2+ab^3 \\ \qquad a^3b+3a^2b^2+3ab^3+b^4 \\ \hline a^4+4a^3b+6a^2b^2+4ab^3+b^4 \end{array}$$

Observe from these examples that the numerical examples in the *expanded forms* (*expansions*) are similar to expansions of $(1+x)^2$, $(1+x)^3$, etc., and this must be so for all powers of the expression.

For example: $(1+x)^3=1+3x+3x^2+x^3$

and $(1+x)^4=1+x(1+x)^3=1+x(1+3x+3x^2+x^3)$

$$
\begin{array}{l}
=1+3x+3x^2+x^3 \\
\quad 1+x \\
\hline
\quad 1+3x+3x^2+x^3 \\
\quad\quad\;\; x+3x^2+3x^3+x^4 \\
\hline
\quad 1+4x+6x^2+4x^3+x^4
\end{array}
$$

In this product take any coefficient, 6 for example. It is obvious that this is the sum of 3, the coefficient of x^2 in the product of $(1+x)^3$, and 3 the coefficient of x in $(1+x)^3$. It will now be seen that the coefficients of x in $(1+x)^4$ may be found from the coefficients of $(1+x)^3$.

Hence $1+3=4$; $3+3=6$; $3+1=4$
and unity as a coefficient at each end.

Now, a *binomial* is an expression consisting of two terms, as $x+y$, and the *binomial theorem* is the rule or formula by means of which any power of a binomial may be found without performing the successive multiplications.

Pascal's Triangle.—The order of the coefficients in $(1+x)^3$ and $(1+x)^4$ will be seen to follow a rule :

1 3 3 1
⋀⋀⋀⋀
1 4 6 4 1

By this method we can thus find, without calculation, the coefficients of the terms in the expansion of $(1+x)^7$ or any other power of $1+x$. Pascal's triangle gives the rule for the coefficients. Thus in $(1+x)^7$ the triangle would be

1 1 $(1+x)^1$
⋀⋀
1 2 1 $(1+x)^2$
⋀⋀⋀
1 3 3 1 $(1+x)^3$
⋀⋀⋀⋀
1 4 6 4 1 $(1+x)^4$
⋀⋀⋀⋀⋀
1 5 10 10 5 1 $(1+x)^5$
⋀⋀⋀⋀⋀⋀
1 6 15 20 15 6 1 $(1+x)^6$
⋀⋀⋀⋀⋀⋀⋀
1 7 21 35 35 21 7 1 $(1+x)^7$

Readers will here see the connection between permutations, combinations, and the binomial theorem.

Thus, the binomial theorem is the formula by which any power of a binomial may be found without performing the successive multiplications. A *binomial* is an algebraic expression consisting of two terms joined by + or —.

Take the product of the n factors of $(1+ay)(1+by)(1+cy)$. . . . It is obvious, at a glance, that

$$(1+ay)(1+by)=1+(a+b)y+aby^2$$

$$\text{and } (1+ay)(1+by)(1+cy)$$
$$=1+(a+b+c)y+(ab+bc+ca)y^2+abcy^3$$

and

$$(1+ay)(1+by)(1+cy)(1+dy) \ldots$$
$$=1+(a+b+c+d+\ \ldots)y$$
$$+(ab+bc+\ \ldots)y^2$$
$$+(abc+bcd+\ \ldots)y^3+\ \ldots$$
$$=1+(\Sigma a)y+(\Sigma ab)y^2+(\Sigma abc)y^3\ \ldots$$

Σ=the summation of the products of ab, etc. We have seen earlier, when dealing with combinations, that all products such as ab are all the combinations possible by selecting two of the letters of the n series, and, as we have seen, they may be designated by ${}^nC_2$.

Therefore: ${}^nC_1$ is the number of terms in Σa
${}^nC_2$ is the number of terms in Σab
${}^nC_3$ is the number of terms in Σabc

and so on.

Summarising: $(1+x)^n$
$$=1+{}^nC_1x+{}^nC_2x^2+C_3x^3\ \ldots\ {}^nC_nx^n$$

This only applies, of course, when the index is a positive integer.

For fractional and negative values of n a different rule applies.

CHAPTER XVIII

ALGEBRAIC FACTORS

I HAVE dealt in an earlier chapter with the elementary algebraic processes of addition, subtraction, multiplication, and division, and it is now necessary for the reader to acquire a knowledge of factors and indices. In any algebraic expression, which is the product of two or more quantities, its *factors* are those quantities.

The factors of $x^2+10x+21$ are $x+3$ and $x+7$, because when multiplied together they produce that expression. The process of finding the factors of an expression is known as *resolution*, or the *resolving* of the expression into its factors. It is a process reverse to that of multiplication. Now in algebra there are many expressions, or *identities*, which occur frequently, and because their factors may easily be recognised they must be memorised. When memorised, the factors of expressions having similar form may be extracted on sight and without calculation. Here they are :

(1) $(a+b)\ (a+b)$, or $(a+b)^2=a^2+2ab+b^2$
(2) $(a-b)\ (a-b)$, or $(a-b)^2=a^2-2ab+b^2$
(3) $(a+b)\ (a-b)=a^2-b^2$
(4) $a^3+b^3\ =(a+b)\ (a^2-ab+b^2)$
(5) $a^3-b^3\ =(a-b)\ (a^2+ab+b^2)$
(6) $(a+b)^3=a^3+3a^2b+3ab^2+b^3$
(7) $(a-b)^3=a^3-3a^2b+3ab^2-b^3$

$a^3+b^3+c^3-3abc=(a+b+c)\ (a^2+b^2+c^2-bc-ca-ab)$.

If we meet the expression $x^2+2xy+y^2$, we know at once that the factors are $(x+y)\ (x+y)$, from the first of the algebraic identities given above ; because these hold true no matter what letters are employed.

From this first identity $(a+b)^2=a^2+2ab+b^2$ we deduce the rule : *The square of the sum of two numbers or quantities is equal to the sum of the squares of the quantities plus twice their product.*

From the second identity $(a-b)^2$ we deduce the rule : *The square of the difference of two numbers or quantities is equal to the sum of the squares of the quantities minus twice their product.*

From the third identity $(a+b)(a-b)$ we deduce the rule : *The product of the sum and the difference of two numbers or quantities is equal to the difference of their squares.*

Suppose we encounter, during calculation, the expression 25^2-24^2. Remembering the rule for the third identity, we write

$$(25+24)(25-24)=49\times 1=49$$

This, it will be agreed, can be solved mentally once the rule is committed to memory. Without knowing the rule, some little time would be taken to work it out.

Another example :

$$\begin{aligned}&\cdot 245^2-\cdot 041^2\\ &=(\cdot 245+\cdot 041)(\cdot 245-\cdot 041)\\ &=\cdot 286\times\cdot 204\\ &=\cdot 058344\end{aligned}$$

Suppose we wish to know the factors of a^4-b^4. Remember ing the third identity,

$$a^4-b^4=(a^2+b^2)(a^2-b^2)$$

We know from the second identity that $a^2-b^2=(a+b)(a-b)$; therefore the factors of

$$a^4-b^4=(a^2+b^2)(a+b)(a-b)$$

Therefore, it can be said that a^n-b^n *can be divided by* $a-b$ *when* n *is odd, and when* n *is an even number it can be divided by* $(a+b)$ *and* $(a-b)$; *and* a^n+b^n *is divisible by* $a+b$ *when* n *is odd.*

Now let us take the expression $a^2+13a+40$, the factors of which are $a+8$ and $a+5$. Here we see that :

The first term is the product of a and a, or a^2.

The middle term is the product of the first term and the sum of 8 and 5.

The last term is the product of 8 and 5.

From this it is a fairly simple matter to find the factors of an expression.

Example : Find the factors of $a^2+8a+15$.

We know that the sum of the two numbers must be 8 and their product 15.

Now 6 and 2, 7 and 1, 5 and 3, 4 and 4 are pairs of numbers which total 8, but of these only one pair, 5 and 3, will give a product of 15. So the two figures required are 5 and 3 and the factors of $a^2+8a+15$ must be $(a+5)$ and $(a+3)$.

Find the factors of $x^2+22x+105$.

The sum of the figures must be 22 and their product 105. Split 22 into pairs of numbers.

21	and	1	15	and	7
20	,,	2	14	,,	8
19	,,	3	13	,,	9
18	,,	4	12	,,	10
17	,,	5	11	,,	11
16	,,	6			

Inspection shows that only one pair of numbers gives the product 105, and those are 7 and 15. Hence the factors of $x^2+22x+105$ are $(x+15)$ and $(x+7)$.

The factors may also be found by *Substitution*, that is, by substituting a value for x to bring the quantity to zero. Take the expression $x^2-15x+56$.

Inspection shows that two possible factors of 56 are -8 and -7, which equal -15 when added together, and $+56$ when multiplied together. To check this assume $x=8$.

Then $64-120+56=0$. As $x=8$,

so $x-8=0$ and therefore -8 must be a factor.

Now try the other factor, 7.

$$49-105+56=0$$

This value of x also satisfies the equation. As x in this case equals 7, then $x-7=0$, and so the factors of $x^2-15x+56$ must be $(x-8)$ and $(x-7)$. Prove by multiplying them together :

$$\begin{array}{l} x-8 \\ x-7 \\ \hline x^2-8x \\ \quad -7x+56 \\ \hline x^2-15x+56 \end{array}$$

Examples.—Factorise the following :

$$a^2-3a-54$$

Answer : $(a+6)\,(a-9)$.

$$a^2-22a+57$$

Answer : $(a-3)\,(a-19)$.

$$x^2-10x+4xy-5y^2-32y+21$$

Answer : $(x-3+5y)\ (x-7-y)$.

$$x^2y+5xy-4x+6y-8$$

Answer : $(xy+3y-4)\ (x+2)$.

$$ax^2y+bx^2y+3ay+3by-az-bz$$

Answer : $(x^2y+3y-z)\ (a+b)$.

$$12x^2-34xy+14y^2$$

Answer : $(3x-7y)\ (4x-2y)$.

$$(a+3b+2c)^2-9(2a+b-c)^2$$

Answer : $(a+3b+2c)^2-(6a+3b-3c)^2$

This is the difference of two squares and

$$x^2-y^2=(x-y)(x+y)$$

Hence the given expression is

$$\begin{aligned}(a+3b+2c-6a-3b+3c)&(a+3b+2c+6a+3b-3c)\\ &=(-5a+5c)(7a+6b-c)\\ &=5(c-a)(7a+6b-c)\end{aligned}$$

In the last example, proceed as follows :

$$\begin{aligned}12x^2+34xy+14y^2&=12\left(x^2-\frac{17}{6}xy+\frac{7}{6}y^2\right)\\ &=12\left\{x^2-\frac{17}{6}xy+\left(\frac{17}{12}y\right)^2+\frac{7}{6}y^2-\left(\frac{17}{12}y\right)^2\right\}\end{aligned}$$

Note : The term $\left(\frac{17}{12}y\right)^2$ is half the coefficient of xy multiplied by y and all squared.

$$\begin{aligned}&=12\left\{\left(x-\frac{17}{12}y\right)^2-\frac{121}{144}y^2\right\}\\ &=12\left\{\left(x-\frac{17}{12}y\right)^2-\left(\frac{11}{12}y\right)^2\right\}\\ &=12\left(x-\frac{17}{12}y+\frac{11}{12}y\right)\left(x-\frac{17}{12}y-\frac{11}{12}y\right)\\ &=12\left(x-\frac{1}{2}y\right)\left(x-\frac{7}{3}y\right)\\ &=3\times4\left(x-\frac{1}{2}y\right)\left(x-\frac{7}{3}y\right)\\ &=(3x-7y)(4x-2y)\end{aligned}$$

What quantity must be added to x^2-2xy to make it divisible by $x-y$?

Answer : y^2.

CHAPTER XIX

Indices

We have seen, in a previous chapter on logarithms, that the number which expresses the *power* of a number is termed the *index*.

Hence in x^2, y^5, a^4, a^2y, xy^7, the *indices* are, respectively, 2, 5, 4, 2, and 7.

The rules regarding indices which have been given in connection with logarithms also apply in algebra. We have seen that, in multiplying two numbers, such as 10^2 and 10^3, we *add* the indices, and write 10^5 as the product. It is important to remember that we must, when multiplying, only add the indices of similar quantities. If we wish to multiply 7^3 by 5^2 we must not add the indices and write 7^5. Simple arithmetic would prove that to be wrong; but if we need to multiply 7^3 by 7^5, we may add the indices, and express the product as 7^8.

The rule is: *To multiply together different powers of the same quantity, add their indices, when the numbers are positive.*

Thus, $x^4 \times x^3 = (x \times x \times x \times x)\ (x \times x \times x) = x^{4+3} = x^7$.

When one power is to be divided by another, the index of the divisor must be subtracted from that of the dividend.

Thus: $$\frac{x^5}{x^2} = \frac{(x \times x \times x \times x \times x)}{(x \times x)} = x^3$$

Therefore: $$\frac{x^5}{x^2} = x^{5-2} = x^3$$

When the indices are symbols, the plus sign must be placed between them when the different powers of the same quantity are to be multiplied together, and the minus sign must be placed between them when they are to be divided.

Thus $x^m \times x^n$ would be written x^{m+n}, and $\frac{x^m}{x^n}$ would be written x^{m-n}.

When a power of a quantity is itself to be raised to a power, the indices must be multiplied together. Thus—

$$(x^2)^5 = x^{2 \times 5} = x^{10}$$

Similarly $\sqrt{x}$ is written $x^{\frac{1}{2}}$

$\sqrt[3]{x}$ is written $x^{\frac{1}{3}}$

$\frac{1}{\sqrt{x}}$ is written $x^{-\frac{1}{2}}$

$\frac{1}{\sqrt[3]{x}}$ is written $x^{-\frac{1}{3}}$

Fractional Indices.—The rules apply to fractional and negative indices. When multiplying a fractional power of a quantity by the fractional power of another quantity the indices are added, and similarly when dividing such fractional powers of quantities the indices are subtracted.

Thus $x^{\frac{1}{5}} \times x^{\frac{1}{3}} = x^{\frac{1}{5}+\frac{1}{3}} = x^{\frac{8}{15}}$ and $x^{\frac{1}{4}} \times x^{\frac{1}{4}} = x^{\frac{1}{4}+\frac{1}{4}} = x^{\frac{1}{2}}$; $x^{\frac{1}{2}} \times x^{\frac{1}{2}} = x^{1}$, and so on.

Hence $x^{\frac{1}{3}}$ means the cube root of x

$x^{\frac{1}{2}}$ means the square root of x

$x^{\frac{2}{3}}$ means the cube root of x^2

also $\frac{x^{\frac{1}{2}}}{x^{\frac{1}{4}}} = x^{\frac{1}{2}-\frac{1}{4}} = x^{\frac{1}{4}}$

and $\frac{x^{\frac{1}{4}}}{x^{\frac{1}{2}}} = x^{\frac{1}{4}-\frac{1}{2}} = x^{-\frac{1}{4}}$

and $\frac{x^{\frac{1}{2}}}{x^{\frac{1}{2}}} = x^{\frac{1}{2}-\frac{1}{2}} = x^{0} = 1$

and $(x^2)^3 = (x^3)^2$ (both equal x^6).

It is most important that the rules of indices be learned, and the following examples should be worked out :

$$x^{1\frac{1}{2}} \times x^{\frac{3}{4}}$$

Answer : $x^{2\frac{1}{4}}$

$$x^{\cdot 125} \times x^{375}$$

Answer : $x^{\cdot 5}$, or $x^{\frac{1}{2}}$, or $\sqrt{x}$

$$\frac{x^5}{x^{\frac{1}{4}}}$$

Answer : $x^{4\frac{3}{4}}$

$$\frac{(x^5)^3}{x^2}$$

Answer : x^{13}

$$x^{\frac{1}{3}} \div x^{\frac{1}{2}}$$

Answer : $x^{-\frac{1}{6}}$

$$\frac{x^{-\frac{1}{2}}}{x^{-\frac{1}{2}}}$$

Answer : $x^0=1$

$$\frac{x^{-\frac{1}{2}}}{x^{\frac{1}{2}}}$$

Answer : x^{-1}, or $\frac{1}{x}$

Notice that the sign of the index of the denominator is changed. Thus, in the last example :

$$\frac{x^{-\frac{1}{2}}}{x^{\frac{1}{2}}}=x^{-\frac{1}{2}-\frac{1}{2}}=x^{-1}, \text{ or } \frac{1}{x}$$

Our knowledge of arithmetical fractions shows that $\frac{x^{-\frac{1}{2}}}{x^{-\frac{1}{2}}}$, $\frac{x^{\frac{1}{2}}}{x^{\frac{1}{2}}}$, $\frac{x^3}{x^3}$, $\frac{x^n}{x^n}$, etc., must all equal 1, because the numerators divide once into the denominators. The expressions are thus similar to $\frac{1}{1}$, $\frac{8}{8}$, $\frac{3{\cdot}75}{3{\cdot}75}$, $\frac{{\cdot}8}{{\cdot}8}$, etc., all of which equal 1.

Another example : $\frac{(x^5)^2}{(x^2)^4}$

$$=x^{5\times2}\div x^{2\times4}$$
$$=x^{10-8}$$
$$=x^2$$

CHAPTER XX

ALGEBRAIC FRACTIONS, H.C.F., AND L.C.M.

Fractions.—Algebraic fractions are treated in the same way as arithmetical fractions, but in algebra it is often more convenient to write out the factors of an expression in simplifying the fractions. Factorisation of algebraic expressions has already been dealt with.

Remember that, as in arithmetic, when comparing, adding, or subtracting fractions, they must have a *common denominator*, and the latter should for ease of working be as small as possible.

Take, for example, $\frac{1}{5a}+\frac{1}{4}$.

Multiply the numerator and denominator of $\frac{1}{4}$ by $5a$, obtaining $\frac{5a}{20a}$; now multiply the numerator and denominator of $\frac{1}{5a}$ by 4, obtaining $\frac{4}{20a}$. Thus $20a$ becomes a common denominator.

Hence $\frac{1}{5a}+\frac{1}{4}=\frac{4}{20a}+\frac{5a}{20a}=\frac{5a+4}{20a}$

Examples.—Simplify $\left(\frac{1}{2}+\frac{1}{2x}\right)\div\left(4x-\frac{4}{x}\right)$

Express this as a fraction in the usual way:

$$\frac{\frac{1}{2}+\frac{1}{2x}}{4x-\frac{4}{x}}$$

Now simplify by reducing to a common denominator:

$$\frac{\frac{x+1}{2x}}{\frac{4x^2-4}{x}}$$

Here it will be seen that the numerator and denominator of $\frac{1}{2}$ have been multiplied by x to produce $\frac{x}{2x}$. As in arithmetic, the value of a fraction is not altered if numerator and denominator are both multiplied or divided by the same number —any convenient number.

Thus, $\frac{x}{2x}+\frac{1}{2x}=\frac{x+1}{2x}$

Now for the denominator portion. Here we write $4x$ as $\frac{4x}{1}$ and multiply numerator and denominator by x, producing $\frac{4x^2}{x}$; so we now have $\frac{4x^2}{x}-\frac{4}{x}$, which is the same as $\frac{4x^2-4}{x}$. Next write the whole as an ordinary fraction:

$$\frac{x+1}{2x}\times\frac{x}{4x^2-4}$$

Note here the application of the rule for division in fractions, which applies: *Invert the divisor and multiply.* It will be seen that $\frac{4x^2-4}{x}$ thus becomes $\frac{x}{4x^2-4}$. Applying the usual rules, the expression becomes:

$$\frac{x(x+1)}{2x(4x^2-4)}$$
$$=\frac{x(x+1)}{4\times2x(x+1)(x-1)}$$
$$=\frac{1}{8(x-1)}$$

Here it will be seen that $x(x+1)$ has cancelled out, leaving 1 for the numerator and $8(x-1)$ as the denominator.

In this example advantage has been taken of factorisation to effect cancellation of quantities common to numerator and denominator.

Highest Common Factor.—It is not always possible, however, to factorise, and we therefore, as in arithmetic, make use of the *Highest Common Factor* (H.C.F.), which is tantamount, as in arithmetic, to finding the *Greatest Common Multiple* (G.C.M.)

The highest common factor of two or more expressions is the highest expression which will divide without remainder into each of the expressions.

Suppose we wish to simplify :

$$\frac{a^4+a^3+2a-4}{a^3+3a^2-4}$$

First we divide the denominator into the numerator thus :

$$\begin{array}{rl} a^3+3a^2-4) & a^4+a^3+2a-4(a-2 \\ & \underline{a^4+3a^3-4a} \\ & \quad -2a^3+6a-4 \\ & \quad \underline{-2a^3-6a^2+8} \\ & \qquad\qquad 6a^2+6a-12 \end{array}$$

This equals $6(a^2+a-2)$.

Now divide a^3+3a^2-4 by a^2+a-2.

$$\begin{array}{rl} a^2+a-2) & a^3+3a^2-4(a+2 \\ & \underline{a^3+a^2-2a} \\ & \quad 2a^2+2a-4 \\ & \quad \underline{2a^2+2a-4} \end{array}$$

The Highest Common Factor is, as shown above, a^2+a-2.

Lowest Common Multiple.—We find the *Lowest Common Multiple* when we wish to compare, subtract, or add two fractions. The lowest common multiple of the denominators of a fraction is the smallest expression into which each of the expressions can be divided without a remainder.

One method is to find the H.C.F. of the expressions in the manner indicated above, divide one of the expressions by it, and then multiply the other expression by the quotient.

In the above example the H.C.F. of the two expressions has been found to be a^2+a-2.

CHAPTER XXI

Quadratic and Cubic Equations

Factors, indices, H.C.F., L.C.M., and fractions are necessary for a proper understanding of methods of solving quadratic equations. As we have seen, any equation in which the highest power of the unknown quantity is 2, such as x^2+y^2, is known as a *quadratic equation*. If the highest power is 3, the equation is a *cubic equation*, and if the highest power is 4, it is a *quartic equation*.

The well-known identity $a^2+2ab+b^2$ is a quadratic expression, and the solution of such equations is concerned with finding their *root*. A quadratic equation may contain the unknown in its first and second powers, or it may contain only the second power of the unknown. The simplest quadratics to solve fall in the latter class, and I shall deal with those first. Let us take a simple example :

$$5x^2=20$$

Then $x^2=\frac{20}{5}=4$

And $x=\sqrt{4}$

$x=\pm 2$ (plus or minus 2)

It is important to remember the rule that *all quadratic equations have two roots*. We have seen in an earlier chapter that when two positive quantities are multiplied together the product is positive, and also that when two negative quantities are multiplied together the result is also positive. When a negative quantity is multiplied by a positive quantity the result is negative. Hence the rule : *The two roots of a positive quadratic are positive* or *negative*.

A real value for a negative quantity cannot be found, and so such quantities are termed *imaginary quantities*, and when roots of quadratics reduce to negative quantities they too are said to be imaginary and without practical meaning.

The square root of any positive number must, therefore, be positive or negative, and although in ordinary arithmetic we express the square root of a number as two positive quantities, they are really positive or negative.

Solve :

$$x=\frac{75}{3x}$$

By cross multiplication $3x^2=75$

$$\text{and } x^2=\frac{75}{3}$$

$$x^2=25$$

$$x=\pm\sqrt{25}$$

$$x=\pm 5$$

Solve :

$$5x^2+125=0$$

$$\therefore\ 5x^2=-125$$

$$x^2=-25$$

$$x=\pm\sqrt{-25}$$

This is an example of an imaginary quantity.

Now, in quadratics of the form $ay^2+by=16$ it is nearly always convenient to equate the quantity to zero as in the above example :

$$ay^2+by-16=0$$

It is necessary to understand what is meant by a *perfect square* before we deal with the general methods and formulæ for the solution of quadratics. For example, the quantity $a^2+10a+25$ is a perfect square because its roots are $\pm(a+5)$. Thus :

$$\sqrt{a^2+10a+25}=\pm(a+5)$$

Also $\sqrt{a^2-8a+16}$ is a perfect square because the factors are $\pm(a-4)$.

From these two expressions we can deduce a rule. Let us take the first two terms of each of the above expressions, a^2+10a and a^2-8a. It will be seen that to convert both to a perfect square 25 has been added to the first, and $+16$ to the second.

It is obvious that $25=\left(\frac{10}{2}\right)^2$, and $16=\left(\frac{8}{2}\right)^2$

This, it will be seen, is in each case the square of half the coefficient of a, and the rule is : *To convert an expression such as x^2+xy into a perfect square, $\left(\frac{x}{2}\right)^2$ (the square of half the coefficient) must be added to it.*

Hence x^2+yx becomes a perfect square, when $\frac{x^2}{4}$ is added to it.

So, in a general quadratic of the form $ay^2+by+c=0$, we must first make the coefficient of y^2 equal to 1 by dividing throughout by a, and the equation becomes

$$y^2+\frac{b}{a}y+\frac{c}{a}=0$$

Transposing :

$$y^2+\frac{b}{a}y=-\frac{c}{a}$$

Now convert the left-hand side into a perfect square by adding the square of half the coefficient of y (applying the rule already given), and to preserve the equality of the equation add the same quantity also to the other side.

The square of half the coefficient of y is $\left(\frac{b}{2a}\right)^2$ and the equation becomes :

$$y^2+\frac{b}{a}y+\left(\frac{b}{2a}\right)^2=-\frac{c}{a}+\left(\frac{b}{2a}\right)^2$$

$$\text{or } y^2+\frac{b}{a}y+\left(\frac{b}{2a}\right)^2=\left(\frac{b}{2a}\right)^2-\frac{c}{a}$$

Clearing the brackets from the right-hand side :

$$y^2+\frac{b}{a}y+\left(\frac{b}{2a}\right)^2=\frac{b^2}{4a^2}--\frac{c}{a}$$

By bringing the right-hand side to the common denominator $4a^2$ and factorising the left-hand side, we have

$$\left(y+\frac{b}{2a}\right)^2=\frac{b^2-4ac}{4a^2}$$

Extract the square root of both sides

$$y+\frac{b}{2a}=\pm\sqrt{\frac{b^2-4ac}{4a^2}}$$

Now transfer $\frac{b}{2a}$ to the other side :

$$y=-\frac{b}{2a}\pm\sqrt{\frac{b^2-4ac}{4a^2}}$$

From which $y=-\frac{b}{2a}\pm\frac{1}{2a}\sqrt{b^2-4ac}$

Here it will be seen that the square root of the denominator $4a^2$ has been placed outside the square-root sign, and :

$$y=\frac{-b\pm\sqrt{b^2-4ac}}{2a}$$

This formula may be generally applied in the solution of quadratic equations. The $\pm$ sign gives two possible solutions :

$$y=\frac{-b-\sqrt{b^2-4ac}}{2a}$$

$$\text{and } y=\frac{-b+\sqrt{b^2-4ac}}{2a}$$

The solutions are, of course, imaginary if the quantities embraced by the square-root sign evolve as negative quantities.

Although two roots are obtained from a quadratic equation only one of them is feasible in relation to practical examples, and the other is ignored.

Examples.—Solve the following quadratic equations :

$$3a^2-3a=16$$

Answer : $a=2{\cdot}863$ or $-1{\cdot}863$

$$\frac{3}{x}-4x=1$$

Answer : $x={\cdot}75$, or -1

$$9x-20=x^2$$

Answer : $x=4$ or 5

In a right-angled triangle the hypotenuse is 9 inches longer than the base, and the perpendicular is 2 inches less than half the base. What are the lengths of the sides ?

Answer : $x^2+\frac{x^2}{4}-2x+4=x^2+18x+81$

$$=\frac{x^2}{4}-20x=77$$

$$x^2-80x=308$$

$$x^2-80x+(40)^2=308+(40)^2 \text{ (adding } \tfrac{1}{2} \text{ coefficient of } x^2)$$

$$(x-40)^2=(43{\cdot}68)^2$$

$$x-40=43{\cdot}68$$

$x=83{\cdot}68$ (Base)

$92{\cdot}68$ (Hypotenuse)

$39{\cdot}84$ (Perpendicular).

Cubic Equations.—A cubic equation is expressed in the standard form :

$$ax^3+bx^2+cx+d=0$$

This is solved in the following manner :

$$x=-\frac{b}{3a}+\left[-\frac{B}{2}+\sqrt{\frac{A^3}{27}+\frac{B^2}{4}}\right]^{\frac{1}{3}}+\left[-\frac{B}{2}-\sqrt{\frac{A^3}{27}+\frac{B^2}{4}}\right]^{\frac{1}{3}}$$

where $A=-\frac{1}{3}\frac{b^2}{a^2}+\frac{c}{a}$

and $B=\frac{2b^3}{27a^3}-\frac{bc}{3a^2}+\frac{d}{a}$

Solve : $2x^3-3x-1=0$

Answer : $\frac{1\pm\sqrt{3}}{2}$, and -1

Solve : $x^3-6x^2+6x+8=0$

Answer : 4, and $1\pm\sqrt{3}$

Solve : $x^3-24x-32=0$

Answer : -4, and $2\pm2\sqrt{3}$

Cubic equations are best solved graphically.

CHAPTER XXII

GRAPHS

If two quantities bear such a relation to one another that a change in one brings about a corresponding change in the other, such are easily represented by means of a *graph*. A graph is drawn on *squared paper*. Squared paper, according to the purpose for which it is to be used, can be obtained with equi-spaced vertical and horizontal lines, of almost any desired spacing. Usually the lines (for convenience in *plotting* decimal quantities) are $\frac{1}{10}$ in., or 1 mm., apart. Sometimes the spaces are $\frac{1}{16}$ in., $\frac{1}{8}$ in., or $\frac{1}{4}$ in. apart. A typical piece of graph paper is reproduced in Fig. 1, showing that the intersecting equi-spaced lines divide the paper into *squares*. The horizontal line OX is the axis of *abscissæ*. The vertical line OY is called the axis of *ordinates*. Every vertical line is an ordinate, and every horizontal line is an *abscissa*.

The space between two ordinates along the line OX may be taken to represent one unit of measurement of one of the quantities, and the space between two abscissæ on the line OY may be taken to represent the unit of the other quantity ; or two or more squares, or a fraction of a square, may represent the unit. It is not necessary to employ the same number of squares to represent a unit of each of the quantities. For example, one square may represent a unit of one of the quantities, and $1\frac{1}{2}$ squares, or two or more squares, may represent a unit of the other. We soon learn how to decide the number of squares to allocate as a unit in each case, in order that the graph may be plotted in the required space.

If we mark along the line OX a distance equal to seven spaces, and along the line OY a distance equal to five spaces, we obtain two points. If we erect a line vertically from the point OX and draw another line horizontally from the point on OY the two lines will intersect. If we deal with a number of observations in the same way we shall obtain a series of such points. These are usually marked with a small × or circle at the points of intersection. It is not, of course, necessary to draw these intersecting lines, as the lines already on the paper act as guides. If these points are connected together by means of a thin piece of wood or celluloid bent to lie along the various points a *curve* is obtained. It may be impossible for the curve to touch all of the points because the line would be irregular or broken. In this case a line which *averages* the points, that is to say, a line which lies evenly among them, may be drawn, and such a line represents the average value of the results plotted. It also serves to indicate errors of plotting or observation. Sometimes a straight line will connect the points, but in graphs such a straight line is still referred to as a curve.

The great advantage of a graph in plotting experimental results is that it enables the relation between two quantities or any change in the relation between those quantities to be seen at once, whereas such changes may not be discerned by ordinary methods of calculation. A simple example will illustrate the great advantage of graphs.

We know that 240 pence make one pound. We can plot a graph of the relation $240=1$, and so be able to ascertain how many pence there are in any number of shillings, or any fraction of a pound. In such a case we should take 20 squares to represent the pound along the line OY, and 24 squares (each square representing ten pennies) along the line OX (Fig. 1). Connecting the two points to the *origin* O, we obtain a curve of the relation between pence and pounds. If, therefore, we desire to know what fraction 180 pence is of one pound we should count 18 spaces along the line OX and note the point where it cuts one of the ordinates.

Equivalent values of relations can be read off direct. The vertical distance along a graph is called the y co-ordinate, and the horizontal distance the x co-ordinate. Suppose, therefore, that we plot two values, say, four spaces along the x co-ordinate and 5 spaces along the y co-ordinate. By joining the point of intersection to the origin in O, a straight line is obtained. When the relation between two variables results in a straight line we are enabled to find what is known as the *equation of the line*. The equation of such a line takes the form :

$$y=ax+b$$

Now a and b are constants, and if the values of x and y are inserted values of a and b can be ascertained. Fig. 2 gives an example. It will be seen that the y co-ordinate and the x co-ordinate intersect at 24 spaces and 40 spaces respectively. Thus, from this graph we can compile values of x and y.

When $x=40,\ y=24$
$x=24,\ y=14{\cdot}4$
$x=16,\ y=\ 9{\cdot}6$

Similarly, we can obtain by direct measurement on the graph any other value of x in relation to y.

Let us insert a pair of these values in the equations :

$$y=ax+b$$

$$24\ =40a+b$$

Subtract : $$14{\cdot}4=24a+b$$

$$9{\cdot}6=16a$$

$$\therefore\quad a=\frac{9{\cdot}6}{16}={\cdot}6$$

Now incorporate this value of a in the equation (given above)

$$24=40a+b$$
$$24=40X\ {\cdot}6+b$$
$$24=4+b$$

from which it is apparent that b is O. Thus the equation of the line is :

$$y={\cdot}6x$$

From this it is obvious that when the equation of a line is of the form $y=ax+b$, then by giving to x the values of 1, 2, 3, 4, . . . the corresponding values of y may be found and the line obtained by plotting.

From a group of values of any two quantities which vary with each other we are able by plotting those values on squared paper to draw a line which *averages the points.* Once the line is drawn we can find its equation, and when the lines of the equation are obtained, then for any value of one of the quantities we are able to obtain the corresponding values for the other, either by calculation from the equation or by inspection of the graph.

Also, if we are given the equation of a line we can assume values for one of the variables, and by calculation find the values of the other. It is always advisable to give definite numerical values to a and b, and to plot the line from them. Next, give a different set of values for a and b, and plot the line again. It will then be noted that the *inclination or slope* of the line depends upon the point at which the line cuts the axis of y on b. In other words, the slope depends upon the term a. Although the terms x and y have been used, other letters can, of course, be used—in fact, are used in some standard equations from which graphs are plotted.

Ohm's Law is a good example, where

$$I=\frac{E}{R}$$

E=Voltage, I=Current, and R=Resistance.

Let us take the simple equation :

$$y=x+2$$

Then $x=0,\ y=2$
$x=1,\ y=3$
$x=2,\ y=4$
$x=3,\ y=5$

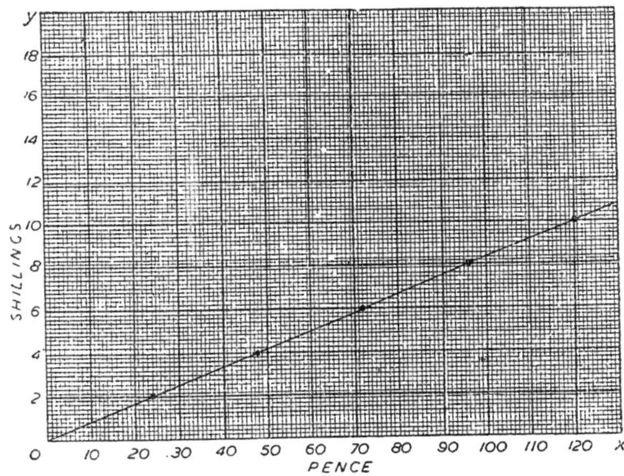

Fig. 1.—Simple graph.

Now let us plot a graph from these values, as in Fig. 3. It will be seen that starting from O the first dot is placed at 2 on the y co-ordinate.

Now taking succeeding values a dot is made at the intersection of 1 on the X co-ordinate with 3 on the OY co-ordinate; at the intersection of 2 with 4 ; and 3 with 5. Connect the points with a line. The line should be drawn as fine as possible so that the graph may be accurately read.

Now let us assume that this line represents the relation of

two variables A and B. We substitute these in the equation previously given. Thus :

$$A=aB+b$$

and in this we substitute the known values of two points in the line. This will yield two equations from which we calculate a and b.

For example, the values along OY may represent values of A, and along OX values of B.

In the graph Fig. 4, when A is 5, B is 3.

When A is 4, B is 2.

We can now substitute these values in the equation :

$$5=a\times3+b$$
$$4=a\times2+b$$
$$1=a$$
$$a=1$$

Substituting this value in the equation $4=2a+b$, $b=2$.

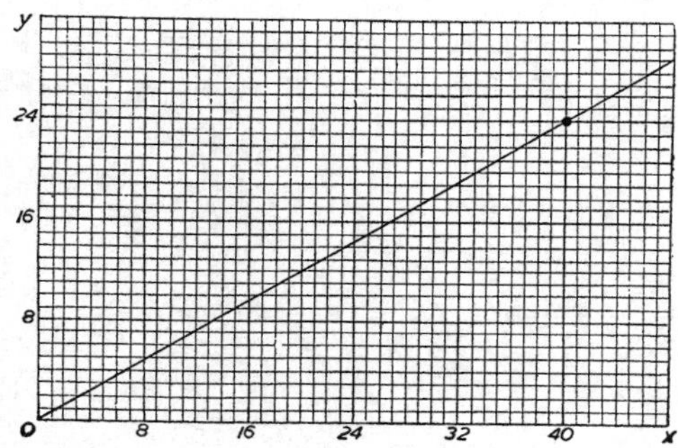

Fig. 2.—Another simple graph.

Now further substitute these values in the equation

$$A=aB+b$$

We obtain $$A=B+2$$

The term b decides the point in the axis of y from which the line is drawn, and by altering its value (letting the value of a remain constant) parallel lines will be obtained.

When the magnitude of a is changed, with b remaining unaltered, a series of lines is obtained which are drawn from the same point, although the slope of each in relation to the axis x will be different.

Now revert to the equation $y=ax+b$, and let $a=1$ and $b=0$, producing $y=x$. When plotted this produces the line xx (Fig. 4). If b is equated to 2 ($b=2$), then $y=x+2$ $y=2$. When this line is plotted it will be parallel to line xx, but two squares above it, as yy.

If we make $a=4$, we shall obtain $y=4x+2$, and this should be plotted.

These are straight line graphs, but all graphs are not straight lines. After plotting the various points and connecting a line

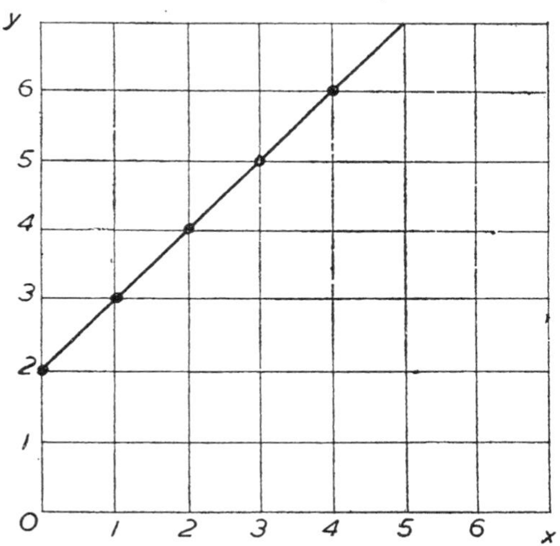

Fig. 3.—Another type of simple graph.

through them, a curved line is often obtained, and it is almost impossible to find a law or equation which will express the relation between the two variables. So we may resort to the artifice of plotting one of the variables, and quantities which are derived from the other, such as the squares, the reciprocals, or the logarithms, to produce a straight line.

Here is a table showing values of x and corresponding values of y :

x	1·0	1·5	2·0	2·5	3·0	3·5	4·0	4·5	5·0	6·0
y	1·5	2·0	2·3	2·7	3·0	3·4	3·7	3·9	4·1	4·55

Fig. 5 is a graph plotted from these values, and it will be seen that the graph produced is a curved line.

It is important, before plotting a graph, to choose a value for each space which will enable the graph to be plotted accurately. The greater the number of squares assigned to each unit of the values more accurate will be the result. At the same time, the number of values to be plotted will to some extent decide the value of each space, so that the graph can be kept within reasonable proportions. Generally, if a small number of values are to be plotted, the number of spaces

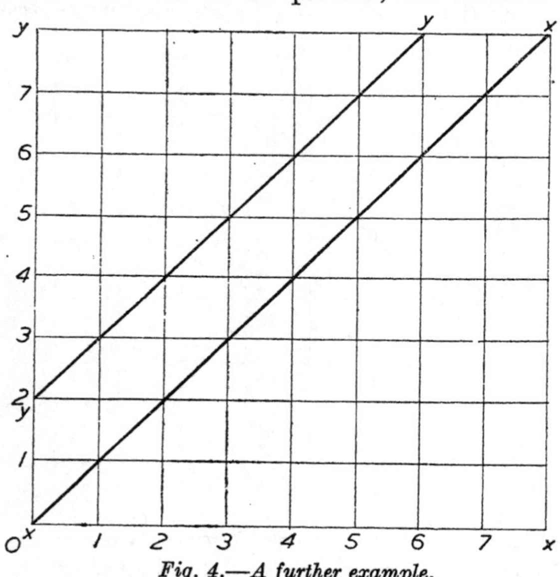

Fig. 4.—A further example.

allotted to each unit may be as large as the graph paper will allow, especially if the values contain decimals, when a space may be given the value of ·1, or 10 spaces may equal ·5. The values or quantities to be plotted will decide the exact value.

From what has been said it will be clear that, having plotted a graph from a series of related quantities, intermediate values of each can be read off.

If we plotted a graph of the squares or the cubes of all numbers from 1 to 10, for example, we could read off the squares or cubes, or the square roots or cube roots of any number within those limits.

It is often convenient to plot the logarithms of one series of quantities. This is the case when plotting the results of a formula which would produce an irregular curve, as already noted. By plotting the logarithms of one series, the resulting graph is approximately a straight line. I shall return to this point later.

Simultaneous equations can conveniently be solved by means of graphs, the algebraic methods having already been given. For example, solve the simultaneous equations by means of a graph

$$2x+3y=12 \quad \ldots\ldots \quad (1)$$
$$6x-3y=12 \quad \ldots\ldots \quad (2)$$

Taking (1) it is obvious that

$$y=\frac{12-2x}{3} \quad \ldots\ldots\ldots \quad (3)$$

and (2) $$y=2x-4 \quad \ldots\ldots\ldots \quad (4)$$

Taking (3), when $x=0$, then $y=4$, and when $x=3$, $y=2$. Plot these two values. Treat equation (4) in a similar way. It will be noted that the two lines intersect, and it is said that this point of intersection is *common to both lines*, and the values of x and y which it indicates are the solution to the two equations (1) and (2). The reader should practise solving some of the simultaneous equations given earlier by means of graphs ; and also plotting curves from any series of related values.

We have seen that, by giving values to x in an equation which expresses the relation between two variables, the corresponding values of y can be found.

When a curve is of the form $y=ax^2$ it is known as a *parabola*, and such a curve can be plotted from the equation :

$$y=\frac{x^2}{4}$$

From which $4y=x^2$ or $y=\frac{x^2}{4}$

Giving values of 1, 2, 3, 4 . . . to x, we shall obtain corresponding values of y. Make a list of the related values :

x	0	1	2	3	4	5
y	0	$\frac{1}{4}$	1	$2\frac{1}{4}$	4	$6\frac{1}{4}$

The curve must thus pass through the origin O of the graph (O being the point of intersection of the two axes).

Now in the chapter on Algebra we saw that when a number,

either positive or negative, is squared, the product is positive. Hence for each value of y there will be *two* values of x, equal in magnitude, but one positive and the other negative, and when plotted these will give two identical curves in reverse, making a combined curve (a parabola) of which half an egg forms a rough example.

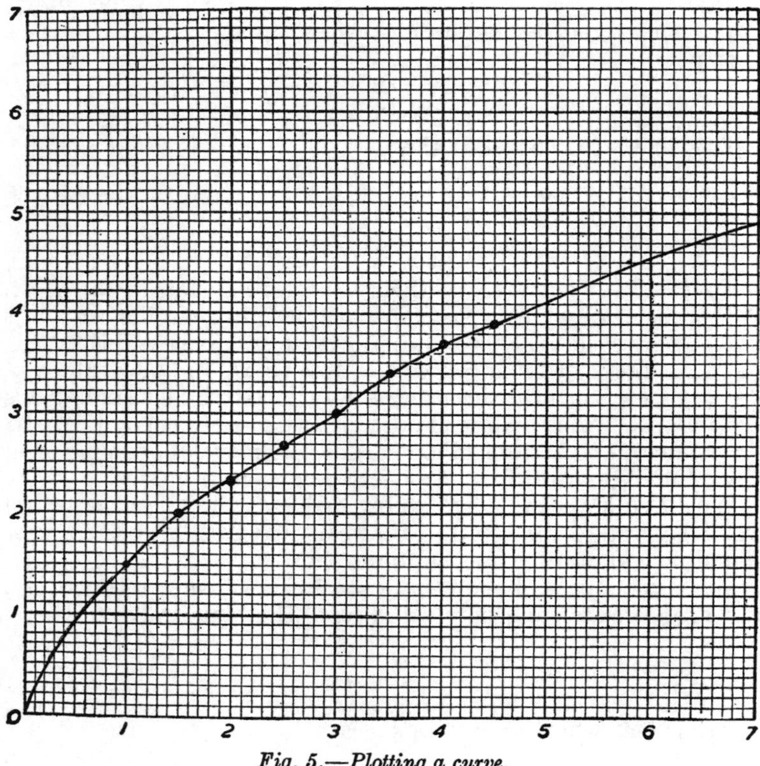

Fig. 5.—Plotting a curve.

It has been assumed that the constant a is positive and$=\frac{1}{4}$; if a is negative and of the same magnitude the equation would be :

$$y=-\tfrac{1}{4}x^2$$

and this will produce another parabola, but inverse to the first and symmetrical to it.

It will be found that the equation $x^2+y^2=20$ (or $x^2+y^2=$ any number other than 0) will produce a circle when plotted (see Fig. 6). Proceed thus :

$$x^2+y^2=20$$
$$y^2=20-x^2$$
$$y=\pm\sqrt{20-x^2}$$

Give values of 0, 1, 2, 3, 4, etc., to x, and calculate the corresponding values of y.

x	0	1	2	3	4	4·47
y	4·47	4·36	4	3·32	2	0

Remember that there will be *two* values of y equal in magnitude, but of opposite sign, and the values of x will similarly be positive and negative. When plotted, the curve produced will be a circle of radius 4·47.

The Hyperbola.—Another curve which frequently occurs in graphs is the *hyperbola*. This curve is such that the difference of the distances from any point on it to two fixed points, the *foci*, is a constant. Such a curve is shown in Fig. 7.

The equation of a rectangular hyperbola is

$$xy=c$$
$$c=\text{a constant}$$

For example, let $xy=10$

$$\therefore\ y=\frac{10}{x}$$

From this, when

$$x=1,\ y=10$$
$$x=2,\ y=5$$
$$x=\tfrac{1}{2},\ y=20$$
$$x=\frac{1}{100},\ y=1000$$

It will be seen that, as the value of x is reduced, the value of y is increased ; when $x=0$, $y=\frac{10}{0}$, and is infinite in value. Hence it will be observed that as the value of x is reduced, the curve gradually approaches the line oy, but it does not cut the axis at any definite point from the origin. The sign for infinity is ∞, and so when $x=0$ $y=\infty$; and when $y=0$ $x=\infty$.

Asymptotes.—In Fig. 7 the lines or axes *ox* and *oy* are said to cut the curve at an infinite distance from the origin *o*. These two lines are called *asymptotes*.

Taking the equation $xy=8$, make a table of corresponding values :

$x=$	0	1	2	3	4	5	6	7	8
$y=$	∞	8	4	$\frac{8}{3}$	2	$\frac{8}{5}$	$\frac{4}{3}$	$\frac{8}{7}$	1

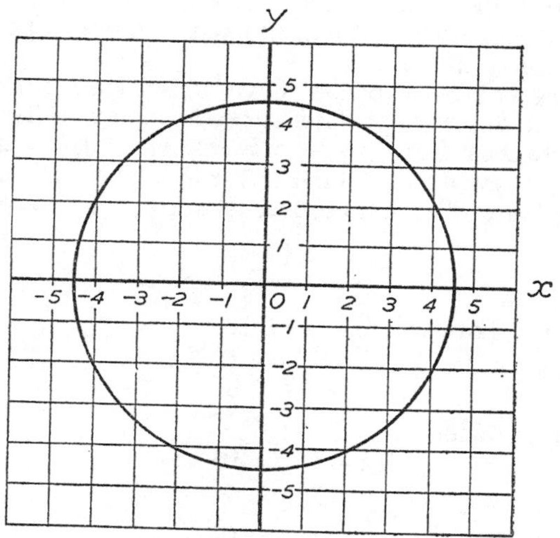

Fig. 6.—Graph of equation $x^2+y^2=20$.

These values are plotted, producing the curve shown in Fig. 7. It is of the greatest importance. It is a *rectangular* hyperbola, and represents the curve of expansion of a gas, or superheated steam.

The constant *c* in this case is found by multiplying pressure and volume together. If pressure be denoted by *p*, and volume by *v*, $c=pv$.

When an equation imposes a law on a point whose motion is represented by the graph, the path traversed by the point is referred to as the *locus*, and the equation of the line or *path* is that which connects the co-ordinates of the point with the law.

It is not possible to give examples of every application of graphs, but some are given to indicate the variety of uses to which they may be put.

When graphs are plotted showing the relation between force and distance, between torque and radian angle, between speed and time, between area and length, and, in fact, between any similar relationships, it is often necessary to calculate the *area under the curve*. Thus, in a graph showing the

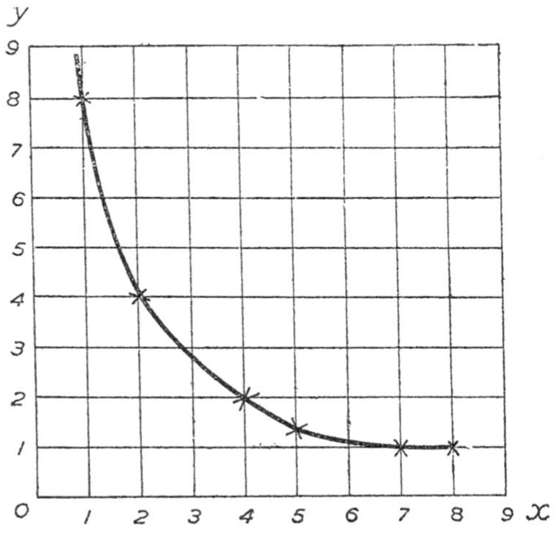

Fig. 7.—Rectangular hyperbola.

relation between force and distance, or between torque and radians, the area under the curve would represent the work done or the energy stored.

If the graph shows the relation between speed and time for a given acceleration, the area beneath the curve indicates distance travelled.

The area may be found by Simpson's *mid-ordinate* rule, sometimes known as the 1 4 2 4 1 rule. In a drawing office an instrument known as a *planimeter* would be used for measuring the area. I shall deal with methods of calculating areas later.

When it is necessary to plot logarithmic values, special graph paper with logarithmic, or semi-logarithmic, rulings may be obtained.

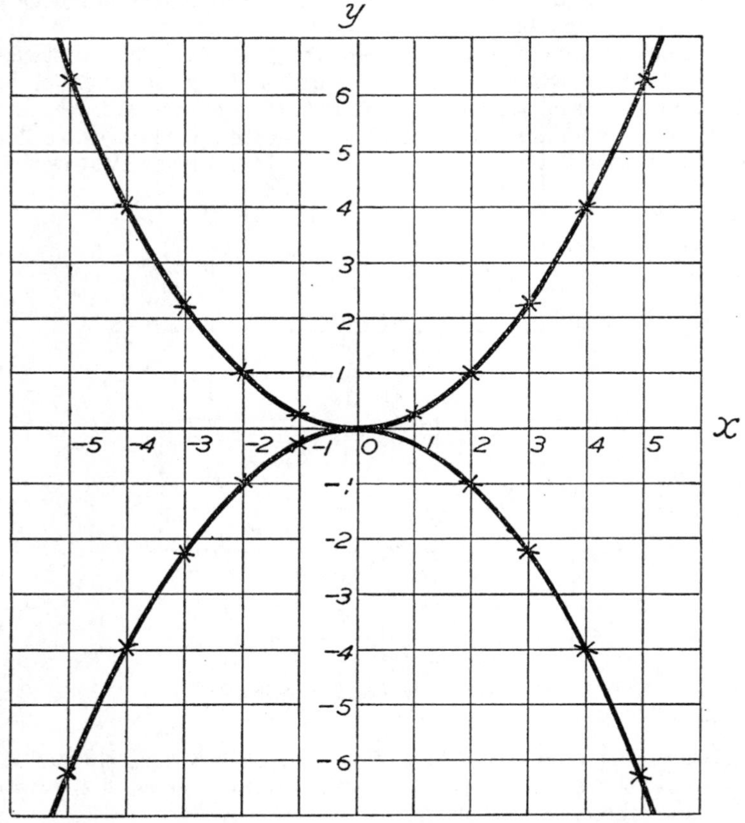

Fig. 8.—The parabola. A graph of the equation $4y-x^2=0$.

It has been shown to be a fairly simple matter, when we have an equation such as $y=5x+16$, to find corresponding values of x and y. In practice, however, we often have to find the equation or law connecting two variables. Fortunately, the law connecting such variables is known in many cases, such as Ohm's law showing relation between current, voltage, and resistance; Hooke's law connecting the tension of a spring with its extension; Boyle's law, and so on. But in the workshop or laboratory we may obtain results from experiments which we must endeavour to correlate, to see if the results obey some law.

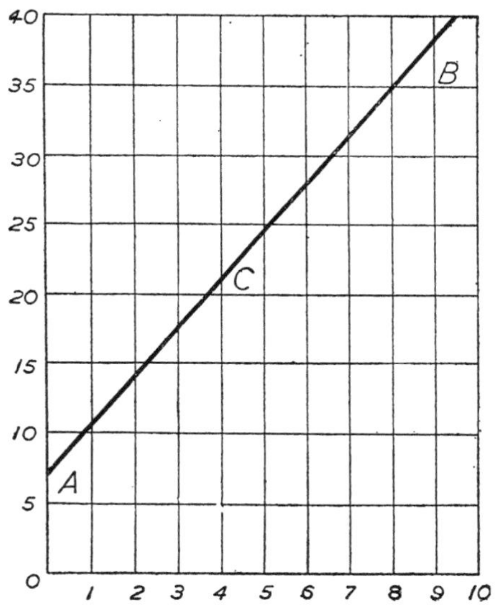

Fig. 9.—Graph of equation $y=7+3{\cdot}5x^2$.

For example, the following results were obtained from tests conducted to show the relation between, say, the load on a bearing and the corresponding friction:

Load $=x$	1	1·5	2	2·3	2·5	2·7	2·8
Friction $=y$	11	15	21·4	25·3	29	32	34·6
$x^2=t$	1	2·25	4	5·29	6·25	7·29	7·84

The assumption was that the relation followed the law $y=a+bx^2$, where $x^2=t$.

Then $y=bt+a$

In the above table it will be seen that the load$=x$, and these values have been squared to produce the values for x^2.

To test the assumption plot a graph of y with corresponding values of t, as in Fig. 9, producing the straight line AB.

Now select any two points on the plotted line, such as BC.

$$\begin{aligned} \text{From the equation}\quad y&=bt+a\\ 21&=4b+a\\ 36&=8\tfrac{1}{3}b+a\\ \text{Subtracting}:\quad 15&=4\tfrac{1}{3}b\\ b&=3{\cdot}5 \text{ approx.}\\ \text{and}\quad a&=7 \end{aligned}$$

As $21=4b+a$, and $b=3{\cdot}5$ the equation becomes :

$$y=7+3{\cdot}5x^2$$

This is the law connecting the two quantities given in the table.

The subject of graphical solution of problems is, of course, a large one, but I have given sufficient examples to refresh the reader's memory of one of the easiest of the mathematical methods of calculation.

CHAPTER XXIII

MENSURATION—TRIGONOMETRY

Terms.—Mensuration is the branch of mathematics which deals with the rules for finding the lengths of lines, the areas of surfaces, and the volumes of solids. There are various subdivisions of this subject.

Geometry is the science of the properties and relations of magnitudes in space of lines, surfaces, and solids.

Longimetry is the measurement of distances.

Planimetry is the measurement of plane surfaces ; in other words, plane geometry as distinct from solid geometry.

Stereometry is the science of measuring solids—solid geometry.

Trigonometry is the branch of mathematics which deals with the measurement of the sides and angles of triangles, and particularly with functions of angles—the sine, cosine, tangent, secant, cosecant, and cotangent.

In order to refresh his memory as to the names of the various geometrical figures, the reader should study the diagrams accompanying this chapter.

Sexagesimals.—We will take trigonometry first. When dealing with angles we make use of the *sexagesimal* system of measurement in which each *degree* is divided into 60 equal parts termed *minutes*, and each minute is further divided into 60 parts termed *seconds*. A *circle* contains 360 degrees or four right angles of 90 degrees. The three angles of any triangle always total 180 degrees.

Degrees are denoted thus : $19°$.
Minutes are denoted thus : $27'$.
Seconds are denoted thus : $35''$.

Although the sexagesimal system is generally employed throughout the world, it has the objection that there are two multipliers, 60 and 90, and for this reason the *centesimal* system was proposed. This is a French system in which the right angle is divided into a hundred parts or *grades*. Each grade consists of a hundred minutes, and each minute one hundred seconds. However, it has not been adopted.

The Radian.—Another system of measurement of angles is used in the higher branches of mathematics, and it makes use of the unit called the *radian* and it is the unit of circular measurement. It is sometimes denoted by 1^c. It will be as well to understand how this unit is derived. Draw a circle of any radius and step off on the circumference a distance equal to the radius. Join these two points to the centre O. The angle AOB (Fig. 10) is the angle (a radian) which is taken as the unit of circular measurement, and in terms of which we measure all others. It is well known that the circumference of a

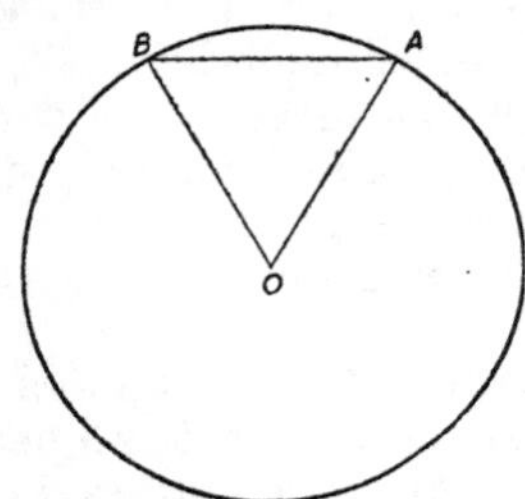

Fig. 10.—Diagram explaining radian.

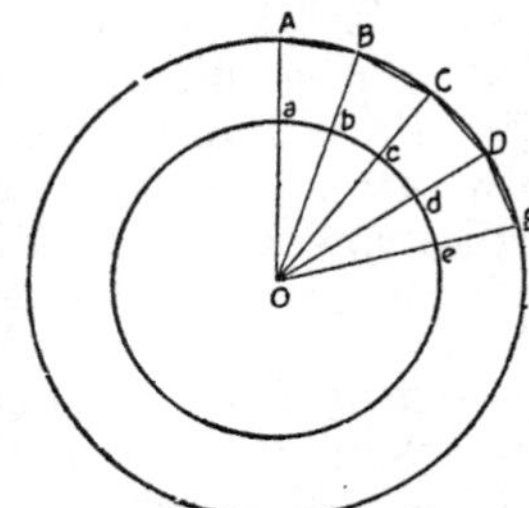

Fig. 11.—Diagram showing relation between radius and circumference.

circle has a constant ratio to its diameter. We can prove this by examining Fig. 11. It will be seen that Oe, Ob, etc., are equal, and that OE and OD are equal, and that the lines ed and ED are parallel.

$$\text{Therefore } \frac{ED}{ed} = \frac{OE}{oe}$$

As the figure shown by Fig. 11, if completed, would form a regular polygon, the length of the perimeter, presuming the polygon to have n sides, will be nED; in other words, the number of sides multiplied by the length of one side.

Therefore we have :

$$\frac{\text{Perimeter of larger polygon}}{\text{Perimeter of smaller polygon}} = \frac{nED}{ned} = \frac{ED}{ed} = \frac{OE}{oe}$$

This must be true irrespective of the number of sides the polygon contains.

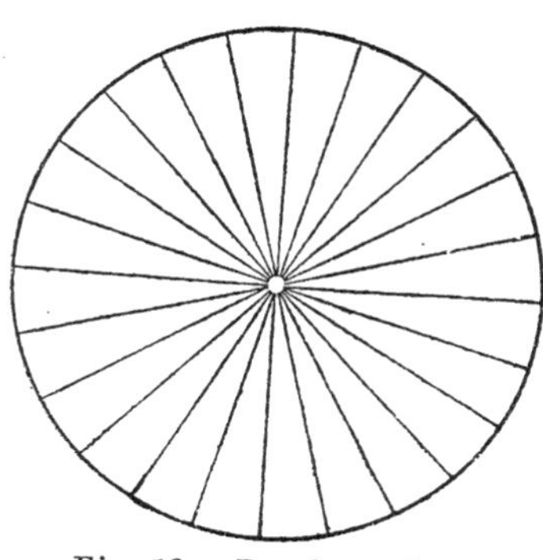

Fig. 12.—Regular polygon.

Now, presume a polygon having an infinitely large number of sides. Then the perimeter would equal the circumference of the circle.

Then,

$$\frac{\text{Circum. of larger circle}}{\text{Circum. of smaller circle}}=\frac{OE}{oe}$$

or

$$\frac{\text{Radius of larger circle}}{\text{Radius of smaller circle}}=\frac{OE}{oe}$$

Therefore :

$$\frac{\text{Circum. of larger circle}}{\text{Radius of larger circle}}=\frac{\text{Circum. of smaller circle}}{\text{Radius of smaller circle}}$$

It is clear from this that the circumference of any circle

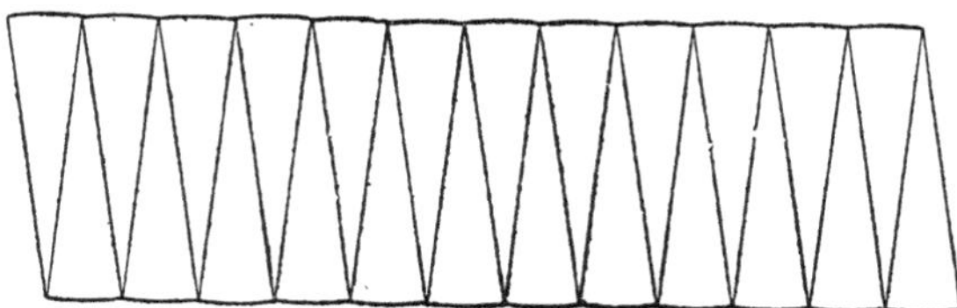

Fig. 13.—The polygon (Fig. 12) rearranged as a rectangle.

divided by its radius is the same for circles of all diameters. In other words, the ratio is constant.

The radian is a constant angle $=\frac{180.}{\pi}$ A right-angle $=\frac{\pi}{2}$ radians.

$$=\frac{180°}{\pi}$$
$$=180°\times{\cdot}3183098862\ .\ .\ .\ .\ .$$
$$=57{\cdot}2957795°$$
$$=57°\ 17'\ 44{\cdot}8''\ \text{approximately}$$

It is obvious, therefore, that a right angle is $\frac{\pi}{2}$ radians

$360^\circ = 4$ right angles $= 2\ \pi$ radians.

Hence, $180^\circ = 2$ right angles $= \pi$ radians.

The number of radians in any angle is equivalent to a fraction having as a numerator the arc which the angle subtends at the centre of any circle, and a denominator which is the radius of that circle.

The Value of Pi.—The value of this constant is always denoted by the Greek letter *pi*, written π. We can prove that *pi* $= 3{\cdot}14159$ approximately. Actually, no one has ever worked out its exact value, although some mathematicians have pursued it to more than 100 decimal places. Let us take a circle 4 in. in diameter, and divide the circumference into 24 equal parts, constructing the polygon shown in Fig. 12. We know that $\pi \times$ diameter equals the circumference, and taking the very approximate value of π to be $3\frac{1}{7}$, the circumference of the circle will be $12\frac{4}{7}$ or $12{\cdot}571$. Now let us consider the polygon as a series of triangles, rearranging them as shown in Fig. 13. If we calculate the length of each piece of the circumference and multiply by 24 the result should be the same as multiplying the diameter by *pi*. The angle contained by each triangular piece will be $\frac{360}{24}$ degrees, or 15 degrees, and, as we shall see later, the length of the arc will therefore be *sin* 15 degrees multiplied by the radius. We consult a table of sines, and find that the sine of 15 degrees is $\cdot 2588$. Multiplying this by the radius (2) we obtain $\cdot 5176$. Multiplying this again by 24 we obtain $12{\cdot}4224$. This compares with $12{\cdot}571$ obtained above. By taking a greater number of triangles, however, so that the curved arc is almost a straight line, we shall obtain a result practically accurate. The difference shown above is accounted for by the fact that the base of each triangle is curved. As I have said, *pi* cannot be worked out exactly, and it is considered as of *incommensurable magnitude*, that is to say, it cannot be exactly calculated. For ordinary calculations $3\frac{1}{7}$ is sufficiently accurate. Other values of *pi* adopted according to the degree of accuracy required in calculations are :

3·142
3·1416
3·14159
3·14159265358979323846

Correct to six decimal places, *pi* equals (by continued fractions) $\frac{355}{113}$, which equals 3·141592. It will be seen that $3\frac{1}{7}$, or $\frac{22}{7}$, equals 3·142857. Thus, this value is only correct to the first two decimal places. Usually the value, 3·14159, is taken for accurate calculation.

Trigonometrical Ratios.—In any right-angle triangle (Fig. 14) the sides vary in relation to one another according to the angle of the opposite side. Thus, the *sine* is obtained by dividing the perpendicular by the hypotenuse. For any given angle this value will hold true no matter what the size of the triangle happens to be. For example, if the perpendicular were 2, and the hypotenuse 3, the sine would equal $\frac{2}{3}$, or ·6666. . . . If we double the size of the triangle, the perpendicular would be 4, and the hypotenuse 6, which still equals ·6666. . . . Obviously, the value of the sine will vary according to the angle, which will control the height of the triangle.

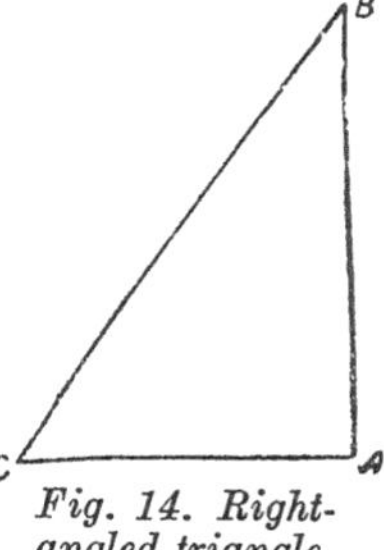

Fig. 14. Right-angled triangle.

Similarly, the cosine of an angle is found by dividing the base by the hypotenuse ; the tangent is found by dividing the perpendicular by the base ; the cotangent by dividing the base by the perpendicular ; the cosecant by dividing the hypotenuse by the perpendicular, and the secant by dividing the hypotenuse by the base.

The amount by which the cosine falls short of unity, that is to say, 1 minus *cos A*, is known as the versed sine, written *versin*, and the amount by which the sine falls short of unity, that is to say, 1 minus *sin A*, is called the coversed sine, written *coversin*.

The abbreviations for all of these terms are as follow :

Sine=sin.
Cosine=cos.
Tangent=tan.

Cotangent = cot.
Cosecant = cosec.
Secant = sec.
Versine = vers. or versin.
Coversed sine = coversin.

It is not often that the two latter ratios are used.

In trigonometry of the circle and hyperbola (hyperbolic trigonometry) the following terms are used :

sinh = hyperbolic sine.
tanh = hyperbolic tangent.
cosh = hyperbolic cosine.

It will be seen from the foregoing that some of the ratios are reciprocals of the others.

Thus :

$$\begin{cases} \textit{Sine} = \dfrac{\text{Perp.}}{\text{Hyp.}} \\ \textit{Cosecant} = \dfrac{\text{Hyp.}}{\text{Perp.}} \end{cases}$$

$$\begin{cases} \textit{Cosine} = \dfrac{\text{Base}}{\text{Hyp.}} \\ \textit{Secant} = \dfrac{\text{Hyp.}}{\text{Base}} \end{cases}$$

$$\begin{cases} \textit{Tangent} = \dfrac{\text{Perp.}}{\text{Base}} \\ \textit{Cotangent} = \dfrac{\text{Base}}{\text{Perp.}} \end{cases}$$

Hence, we may write :

$$\textit{Cotangent} = \frac{1}{\textit{tangent}}$$

$$\textit{Secant} = \frac{1}{\textit{cosine}}$$

$$\textit{Cosecant} = \frac{1}{\textit{sine}}$$

There are many other ratios which may be deduced and which are given later.

Evaluating the above, by cross multiplication :

$$\textit{Cotan.} \times \textit{tan.} = 1$$
$$\textit{Sec.} \times \textit{cos.} = 1$$
$$\textit{Cosec.} \times \textit{sine.} = 1$$

Therefore, if one of the trigonometrical ratios of an angle be known, we can calculate the other ratios.

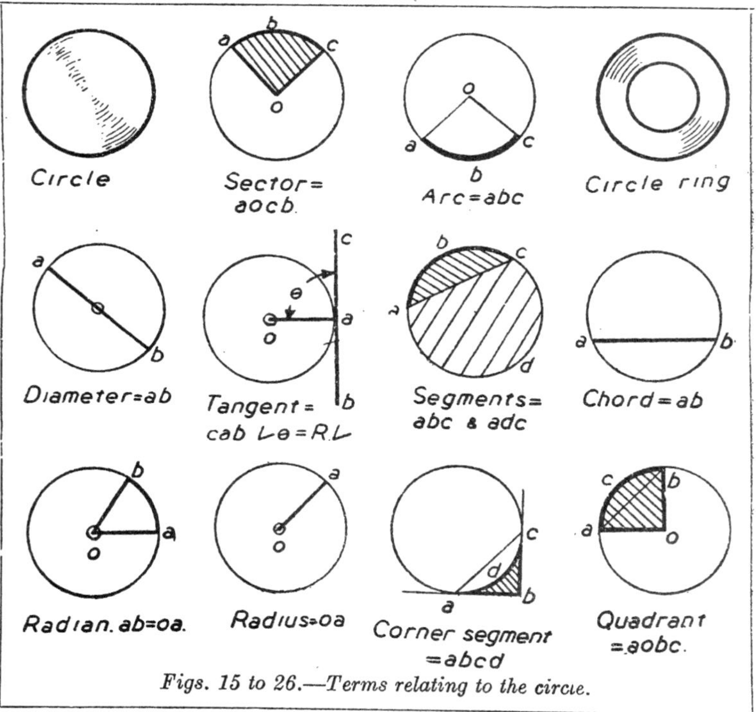

Figs. 15 to 26.—Terms relating to the circle.

Pythagoras' Rule.—The rule relating to right-angled triangles is : The square of the base, plus the square of the height, is equal to the square of the hypotenuse.

As the square of the base added to the square of the height equals the square of the hypotenuse, it follows that if we know the length of two sides of a right-angled triangle we can calculate the length of the third side. In Fig. 69 let the length of the base be 3 in., and the height 4 in. Then :

$$\begin{aligned} \text{Hypotenuse} &= \sqrt{3^2+4^2} \\ &= \sqrt{9+16} \\ &= \sqrt{25} \\ &= 5 \end{aligned}$$

If we were given the lengths of the hypotenuse and one other side we *subtract* from the square of the hypotenuse the square of the other known side, and the *square root* of the answer

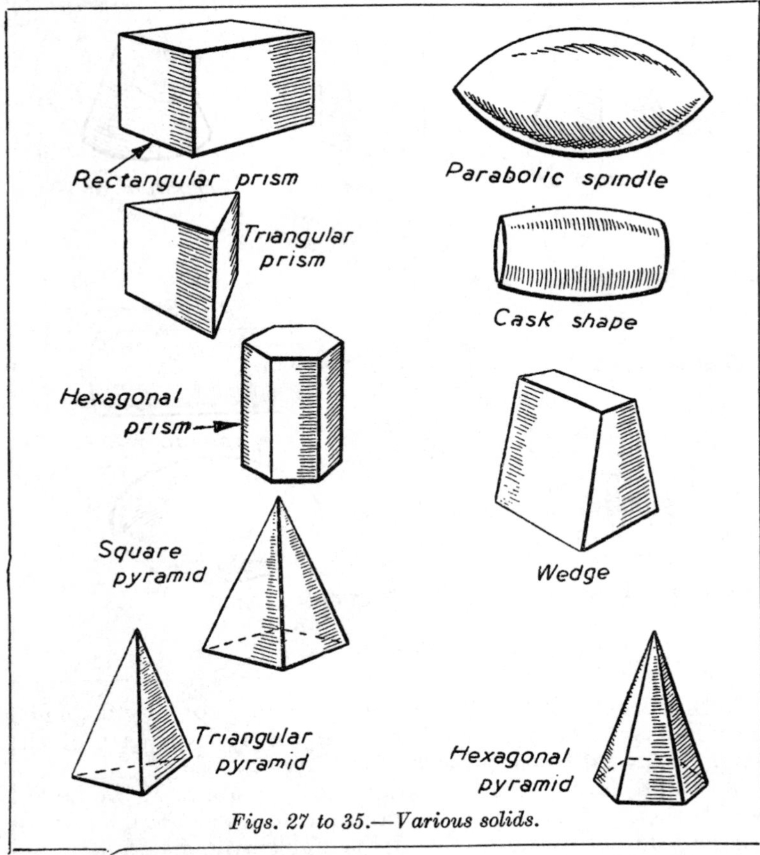

Figs. 27 to 35.—Various solids.

the length of the third side. If the two known sides in the example given are the hypotenuse and the base :

$$5^2 - 3^2 = \text{height}^2$$
$$25 - 9 = \text{height}^2$$
$$\text{height} = \sqrt{16}$$
$$= 4$$

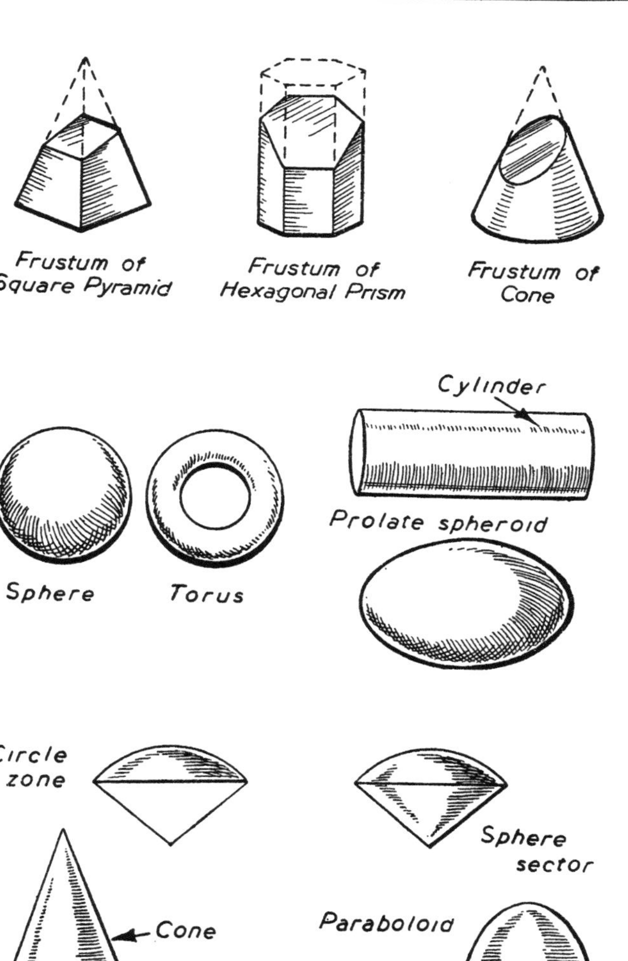

Figs. 36 to 46.—Further solids.

Similarly, knowing the hypotenuse and the height:

$$5^2-4^2=\text{base}^2$$
$$25-16=\text{base}^2$$
$$\text{base}=\sqrt{9}$$
$$=3$$

Often, however, we are given the angle and the length of

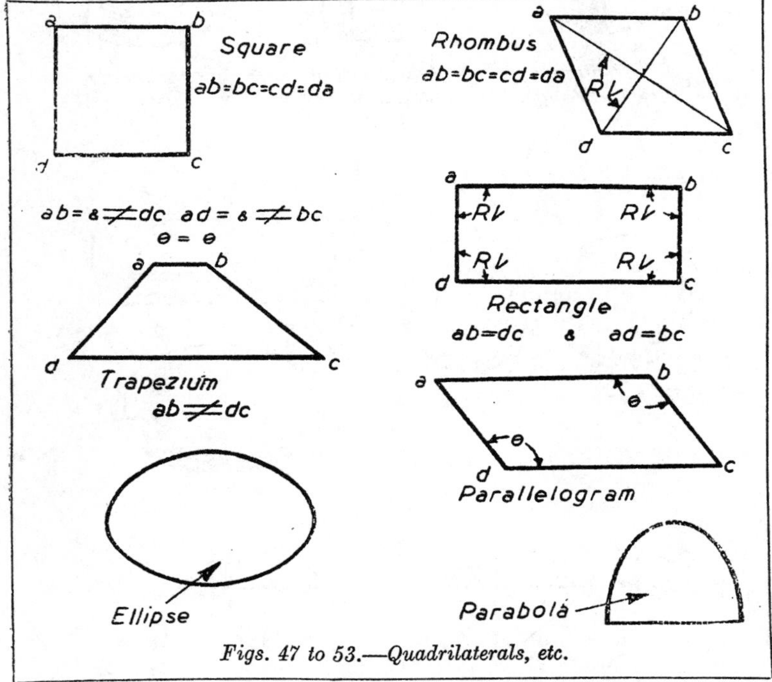

Figs. 47 to 53.—Quadrilaterals, etc.

one side, and we then make use of the trigonometrical ratios—sine, cosine, tangent, etc.

For example, we are given the angle A as 60 degrees, Fig. 70, and the length of the base as 3 in. We have seen that

$$\tan a=\frac{\text{Perp.}}{\text{Base}}$$

or

$$\tan a=\frac{BC}{3}$$

Consult a table of *natural tangents*, and read off the tangent of 60 degrees. It is found to be 1·7321. Therefore :

$$1{\cdot}7321=\frac{BC}{3}$$

By cross multiplication :

$$\begin{aligned} BC&=3\times 1{\cdot}7321\\ &=5{\cdot}1963 \end{aligned}$$

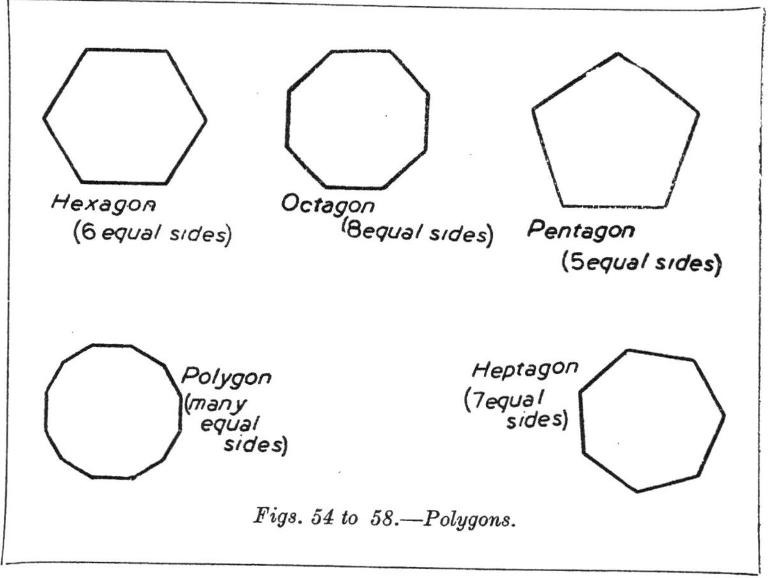

Figs. 54 to 58.—Polygons.

Similarly, if we were given the perpendicular, and the angle as 60 degrees :

$$1{\cdot}7321=\frac{5{\cdot}1963}{AB}$$

Hence :

$$\begin{aligned} AB\times 1{\cdot}7321&=5{\cdot}1963\\ AB&=\frac{5{\cdot}1963}{1{\cdot}7321}\\ &=3 \end{aligned}$$

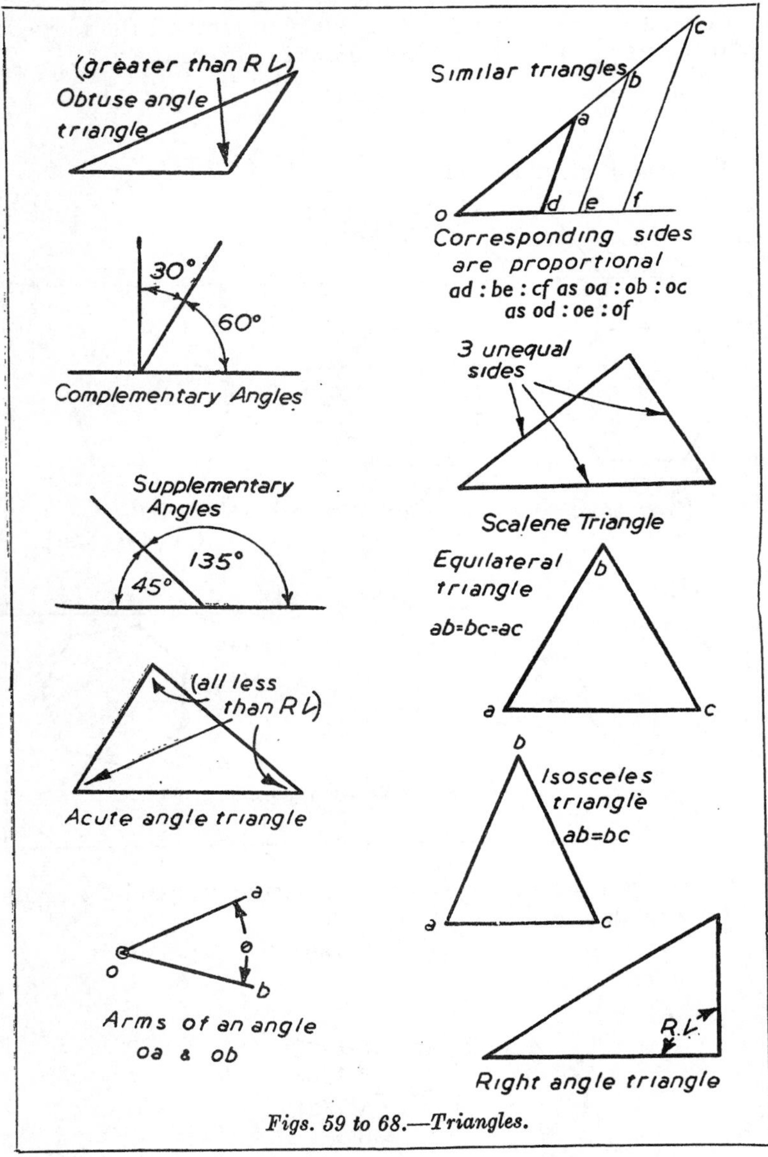
(greater than R ∟)
Obtuse angle
triangle
30°
60°
Complementary Angles
Supplementary
Angles
135°
45°
(all less
than R ∟)
Acute angle triangle
a
o
b
Arms of an angle
oa & ob
Similar triangles
c
b
a
o
d
e
f
Corresponding sides
are proportional
ad : be : cf as oa : ob : oc
as od : oe : of
3 unequal
sides
Scalene Triangle
Equilateral
triangle
ab=bc=ac
b
a
c
b
Isosceles
triangle
ab=bc
a
c
R.∟
Right angle triangle

Figs. 59 to 68.—Triangles.

More conveniently, it is always wise to select a function which places the unknown side as a numerator. Thus, given the angle and the base, we should select the cotangent, which is, as we have seen, $\frac{\text{Base}}{\text{Perp.}}$

Remembering that the cotangent has the same value as the

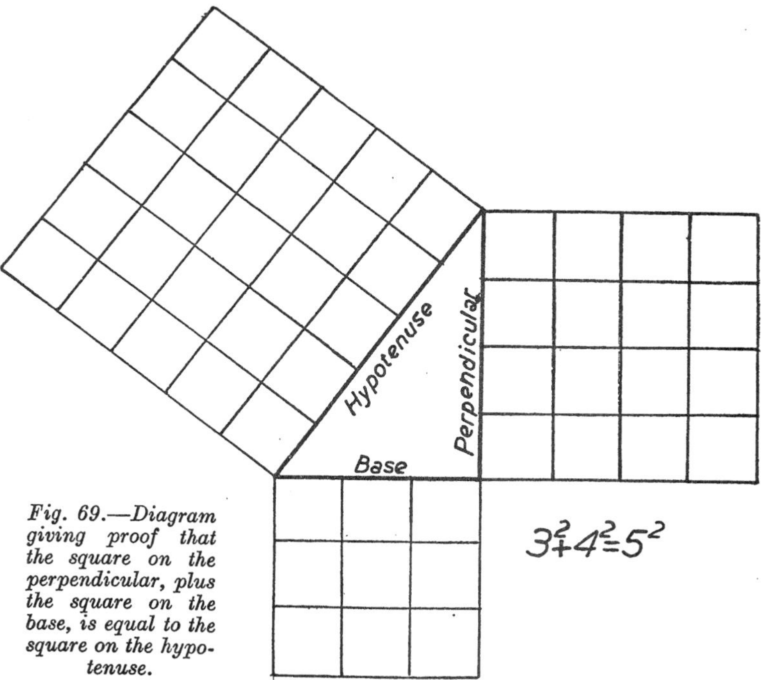

Fig. 69.—Diagram giving proof that the square on the perpendicular, plus the square on the base, is equal to the square on the hypotenuse.

tangent of the complementary angle, it is obvious that cotan 60 degrees is the same as tan 30 degrees (in other words, 90 degrees—60 degrees), and, by consulting the tables, we find that tan 30 degrees is ·5774. Therefore :

$$\cdot 5774 = \frac{AB}{5\cdot 1963}$$

$$AB = \cdot 5774 \times 5\cdot 1963$$

$$= 3\cdot 000$$

If we were given the angle and the perpendicular, and required to find the hypotenuse, we should use the cosecant of the angle; if the hypotenuse and angle were given and required to find the perpendicular we should use the sine; if the hypotenuse and angle were given and we required to find the base we should use the cosine; and if the base and angle were given and we required to find the hypotenuse, we should use the secant.

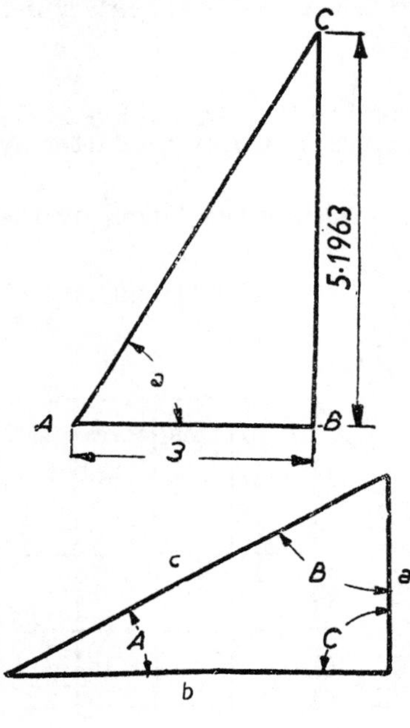

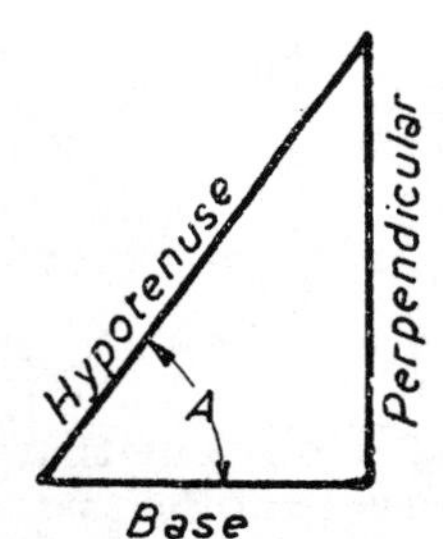

Figs. 70 to 72.—Right-angled triangles.

As the sine and cosecant are reciprocals,

$$\text{cosec } A^\circ = \frac{1}{\sin 60^\circ}$$

As the cosine and secant are reciprocals,

$$\cos A^\circ = \frac{1}{\sec 60^\circ}$$

As the tangent and cotangent are reciprocals,

$$\tan A^\circ = \frac{1}{\cot 60^\circ}$$

Thus, the sine of any angle is the same as the cosine of 90 degrees minus that angle. The cosine of any angle is the same as the sine of 90 degrees minus that angle. The tangent of any angle is the same as the cotangent of 90 degrees minus that angle. These rules hold in the converse—that is to say:

$$\text{cosec } A = \sin (90^\circ - A)$$
$$\sec A = \cos (90^\circ - A)$$
$$\cot A = \tan (90^\circ - A)$$

Now, memorise the following rules :

Given the angle and the hypotenuse, multiply the latter by the sine to obtain the perpendicular.

Given the angle and the perpendicular, multiply the latter by the cosecant to obtain the hypotenuse.

Given the angle and the hypotenuse, multiply the latter by the cosine to obtain the base.

Given the angle and the base, multiply the latter by the secant to obtain the hypotenuse.

Given the angle and the base, multiply the latter by the tangent to obtain the perpendicular.

Given the angle and the perpendicular, multiply the latter by the cotangent to obtain the base.

We can now construct a table, and in this connection reference should be made to Fig. 71.

Parts Given.	Parts to be Found (see Fig. 71).				
	A	B	a	b	c
a and c	$\sin A=\frac{a}{c}$	$\cos B=\frac{a}{c}$		$b=\sqrt{c^2-a^2}$	
a and b	$\tan A=\frac{a}{b}$	$\cot B=\frac{a}{b}$			$c=\sqrt{a^2+b^2}$
c and b	$\cos A=\frac{b}{c}$	$\sin B=\frac{b}{c}$	$a=\sqrt{c^2-b^2}$		
A and a		$B=90°-A$		$b=a\times\cot A$	$c=\frac{a}{\sin A}$
A and b		$B=90°-A$	$a=b\times\tan A$		$c=\frac{b}{\cos A}$
A and c		$B=90°-A$	$a=c\times\sin A$	$b=c\times\cos A$	

There are many standard trigonometrical formulæ which are derived from the standard functions of angles. I give the more important of them here :

$$\tan 45° = 1$$

$$\sin^2\theta + \cos^2\theta = 1$$

$$\sec^2\theta = 1 + \tan^2\theta$$

$$\text{cosec}^2\theta = 1 + \text{cotan}^2\theta$$

$$\text{vers}\,\theta = 1 - \cos\theta$$

$$\sec\theta = 1 \div \cos\theta$$

$$\text{coversin}\,\theta = 1 - \sin$$

$$\text{cotangent}\,\theta = 1 \div \tan$$

$$\text{cotangent}\,\theta = \cos \div \sin$$

$$\sin 0° = 0$$

$$\cos 0° = 1$$

$$\sin 30° = \frac{1}{2}$$

$$\cos 60° = \frac{1}{2}$$

$$\cos 30° = \frac{\sqrt{3}}{2}$$

$$\sin 45° = \cos 45° = \frac{1}{\sqrt{2}}$$

$$\sin 60° = \frac{\sqrt{3}}{2}$$

$$\sin 90° = 1$$

$$\cos 90° = 0$$

$$\sin 15° = \frac{\sqrt{3}-1}{2\sqrt{2}}$$

$$\cos 15° = \frac{\sqrt{3}+1}{2\sqrt{2}}$$

$$\tan\theta \times \text{cotan}\,\theta = 1$$

Complementary Angles.—

$$\cos\theta=\sin(90-\theta)$$
$$\sin\theta=\cos(90-\theta)$$
$$\tan\theta=\cot(90-\theta)$$
$$\sec\theta=\operatorname{cosec}(90-\theta)$$
$$\cot\theta=\tan(90-\theta)$$
$$\operatorname{cosec}\theta=\sec(90-\theta)$$

Angles between 90° and 180°.—When θ is between 90° and 180°:

$$\cot\theta=-\cot(180-\theta)$$
$$\tan\theta=-\tan(180-\theta)$$
$$\sin\theta=\sin(180-\theta)$$
$$\sec\theta=-\sec(180-\theta)$$
$$\operatorname{cosec}\theta=\operatorname{cosec}(180-\theta)$$
$$\cos\theta=-\cos(180-\theta)$$

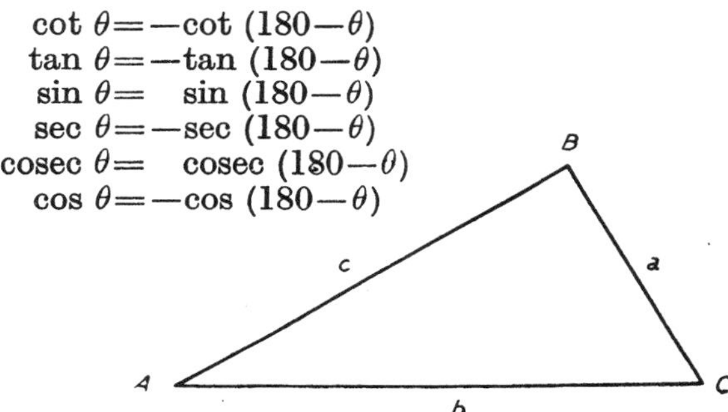

Fig. 72a.—Diagram relating to the Sine Rule.

Sine Rule.—In any triangle

$$\frac{a}{\sin A}=\frac{b}{\sin B}=\frac{c}{\sin C}$$

Here are some examples which the reader should work out, and verify the answers. The examples all relate to Fig. 72.

The angle A is 27 degrees and the base is 3·25. Find the hypotenuse (remember that the secant$=1\div\cos$). Answer: 3·6476.

Angle A is 47 degrees and the perpendicular is 4·45. Find the base. Answer: 4·1496.

Angle A is 64 degrees, and hypotenuse is 7·38. Find the perpendicular. Answer: 6·6331.

Angle A is 31 degrees, and the base is 2·18. Find the perpendicular. Answer: 1·3099.

Angle A is 71 degrees, and the perpendicular is 3·74. Find the hypotenuse (remember cosecant$=1\div\sin$). Answer: 3·955.

These examples should be varied, and the other sides of the triangle calculated, to check the previous answers.

CHAPTER XXIV

Areas of Circle, Triangles, and Quadrilaterals

The reader after a little practice should be able to calculate the lengths of sides of triangles and to calculate angles. As every triangle contains 180 degrees it is a simple matter in a right-angle triangle, where one of the angles is 90 degrees

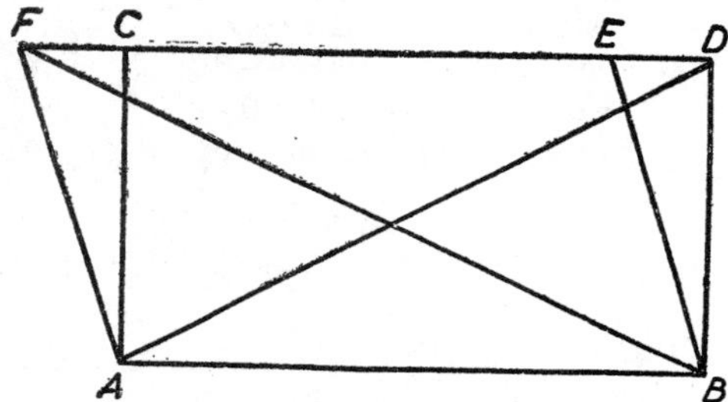

Fig. 73.—Diagram illustrating area of triangle.

and the other is known, to calculate the third angle. For example, if one of the other angles is 30°, 90°+30°=120°, and 180°−120° gives the third angle as 60 degrees.

Area of Triangles.—The various forms of triangles and quadrilaterals have already been given. First draw the rectangle *ABCD* (Fig. 73), and then construct on it the parallelogram *ABEF*. It is obvious that the area of the rectangle and of the parallelogram are equal because parallelograms on the same base and of the same *altitude* must equal one another. It is also obvious that the triangle *ABD* is one-half the area of the rectangle *ABCD*. It is also apparent that the triangle *ABF* is one-half the area of the parallelogram *ABEF*. It follows, therefore, that the two triangles *ABD* and *ABF* must also be equal in area. Expressed as a rule : *The area is equal to half the product of the base and the altitude in each case.* Expressed in the more usual way :

$$\text{Area of triangle} = \tfrac{1}{2}(\text{base} \times \text{altitude})$$
$$= \tfrac{1}{2}ab$$

Thus, if we know two sides of a right-angle triangle, we can find the area. For example, a right-angle triangle has sides of lengths 6, 8, and 10 ft. respectively. Obviously the longer side must be the hypotenuse, so we ignore that and multiply 6 and 8 together, dividing the result by 2. The area of the triangle will therefore be :

$$\tfrac{1}{2}(6\times 8)$$
$$=\frac{48}{2}=24 \text{ sq. ft.}$$

If a triangle has two equal sides, then the angles opposite those sides are equal and it is an isosceles triangle.

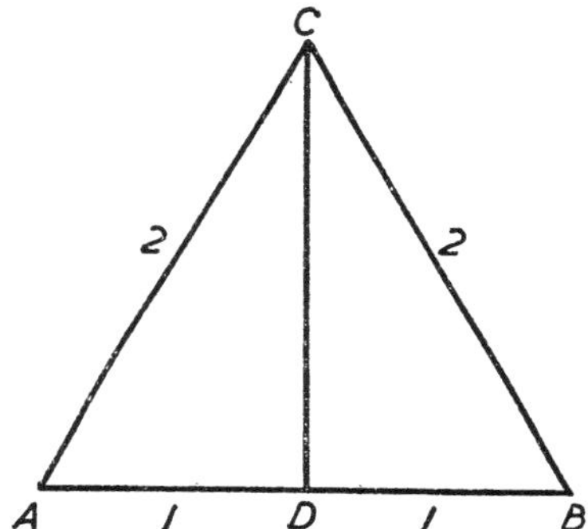

Fig. 74.—Area of equilateral triangle.

An *equilateral* triangle has sides of equal length, and therefore each angle must be 60 degrees, that is to say, 180 degrees divided by 3.

If each side of the equilateral triangle is 2 in. long the area will be $AD\times$CD. We must calculate CD by Pythagoras' Rule (Fig. 74).

$$AD^2+CD^2=AC^2$$
$$CD=\sqrt{2^2-1^2}$$
$$=\sqrt{3}$$

Therefore, *when one angle of a right-angle triangle is* 60 *degrees the three sides are in the ratio of* 2, 1, and $\sqrt{3}$ *to one another.* Hence, the area of the triangle $ABC=\frac{1}{2}\times 2\times\sqrt{3}=\sqrt{3}$. We see, hence, that to calculate the area of an equilateral triangle *we must know the vertical height.*

We can also calculate the area of a triangle if we know the length of the three sides by applying the formula :

$$\text{Area}=\sqrt{s(s-a)(s-b)(s-c)}$$
$$\text{Where } s=\frac{a+b+c}{2}$$

and a, b, c are the lengths of the sides respectively. In other words, *the length of each side is subtracted from half the sum of the sides and the three remainders are multiplied together. The result is multiplied again by half the sum of the sides, and the square*

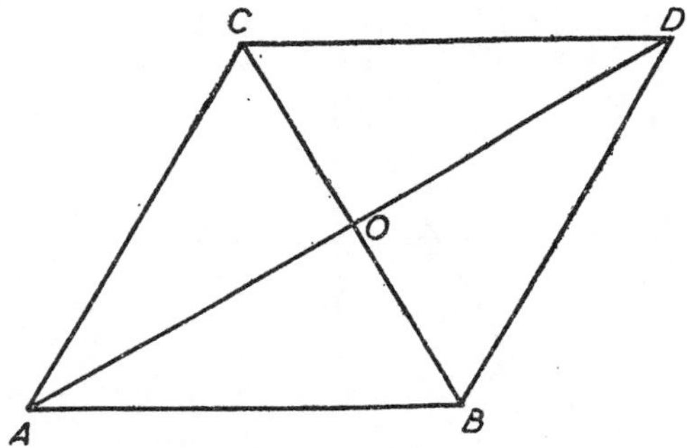

Fig. 75.—Area of rhombus.

root of the product is extracted to obtain the area. This rule also applies to right-angle triangles.

In the right-angle triangle dealt with in the previous example, in which the sides were 6 ft., 8 ft., and 10 ft., respectively,

$$s=\frac{6+8+10}{2}=12$$
$$\text{Area}=\sqrt{12(12-6)(12-8)(12-10)}$$
$$=\sqrt{12(6)(4)(2)}$$
$$=\sqrt{576}$$
$$=24 \text{ sq. ft.}$$

Area of Rhombus.—It is important to remember, as will be seen from Fig. 75, that the *diagonals of a rhombus intersect one another at right angles*. It will be observed from this illustration that OB represents the height of the triangle ABD, and OC the height of the triangle ABD, and OC the height of the triangle

ADC. From this we deduce that *the area of a rhombus is equal to the area of the two triangles.* This equals :

$$\tfrac{1}{2}(AD\times OB)+\tfrac{1}{2}(AD\times OC)$$
$$=\tfrac{1}{2}(AD\times BC)$$

It will be seen, therefore, that *the area of a rhombus is equal to half the product of the two diagonals.* It is important to remember that *all the sides of a rhombus are equal in length.* In examinations it is usual to give the length of a diagonal and the length of one side, from which the student is asked to calculate the area. This is done using the rule given previously, calculating the area of the triangle ABD and multiplying by 2.

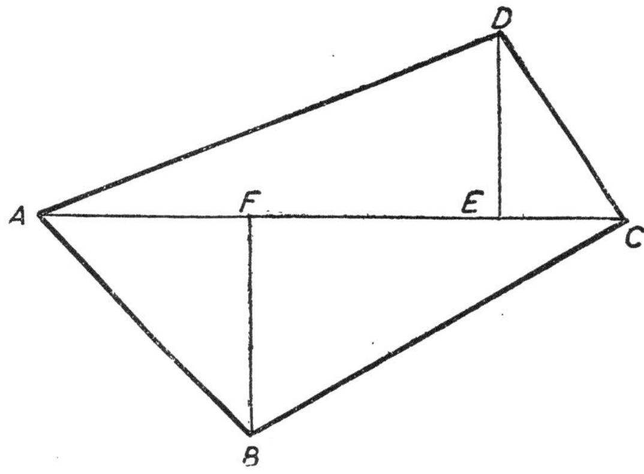

Fig. 76.—Area of quadrilateral.

Area of Quadrilaterals.—All four-sided figures are known as *quadrilaterals* whatever the lengths of their sides (Fig. 76). Divide the figure into two triangles ; obviously the area of the quadrilateral is the sum of the areas of the two triangles. Perpendicular to AC erect the lines FB and ED, thus creating right-angle triangles ABF and ECD. Obviously the area of the triangle $ACD=\tfrac{1}{2}(AC\times DE)$, and the area of $ABC=\tfrac{1}{2}(AC\times FB)$. It follows that the area of the quadrilateral $=\tfrac{1}{2}AC(FB+DE)$.

Reduced to a rule, *the area of the quadrilateral is equal to half the product of one of the diagonals and the sum of the two perpendiculars.*

Polygons.—A figure having more than four sides is known

as a *polygon*, and it is well to learn the correct names of figures having various numbers of sides.

A three-sided figure is a *triangle*.

A four-sided figure (*square*, *rectangle*, *parallelogram*, *trapezium*, *rhombus*) is known as a *quadrilateral*.

A *rectangle* is a four-sided figure containing four right angles.

A five-sided figure is known as a *pentagon*.

A six-sided figure is known as a *hexagon*.

A seven-sided figure is known as a *septagon*, *septangle* or *heptagon*.

An eight-sided figure is known as an *octagon*.

A nine-sided figure is known as a *nonagon*.

A ten-sided figure is known as a *decagon*.

An eleven-sided figure is known as an *undecagon*, *endecagon* or *hendecagon*.

A twelve-sided figure is known as a *duodecagon* or *dodecagon*.

A fifteen-sided figure is known as a *quindecagon*.

Two lines which intersect one another are known as an *angle*. A six-sided figure as shown in Fig. 77 contains six

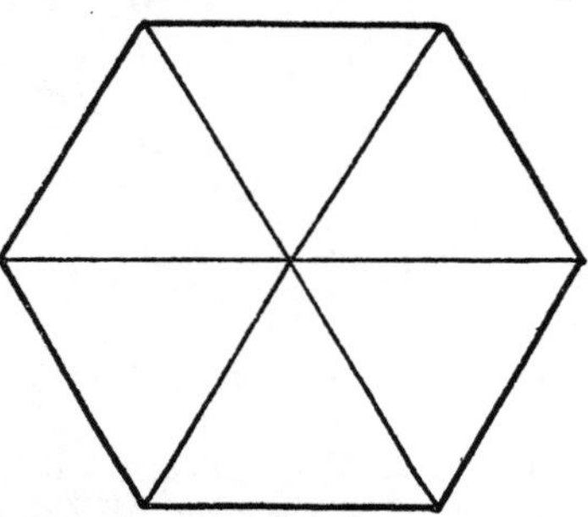

Fig. 77.—Area of regular hexagon.

equal equilateral triangles, and thus the figure is a *regular hexagon*. Obviously the area of the figure is six times the area of one triangle. It will be found from previous reasoning that the area of a hexagon

$$=6\times\frac{a^2}{4}\sqrt{3}$$

$$=\frac{3}{2}a^2\sqrt{3}$$

where a=length of one side.

A shorter rule for the area of a hexagon is: 2·598 (Length of side)^2.

It follows that if we know the area of a hexagon we can find the length of the side. For example, if the area of a hexagon is 480 sq. ft. we have :

$$480 = 2{\cdot}598a^2.$$

$$\text{Therefore } a^2 = \frac{480}{2{\cdot}598}$$

$$a = \sqrt{\frac{480}{2{\cdot}598}}$$

$$a = \sqrt{180{\cdot}47}$$

$$a = 13{\cdot}44 \text{ ft. (approx.).}$$

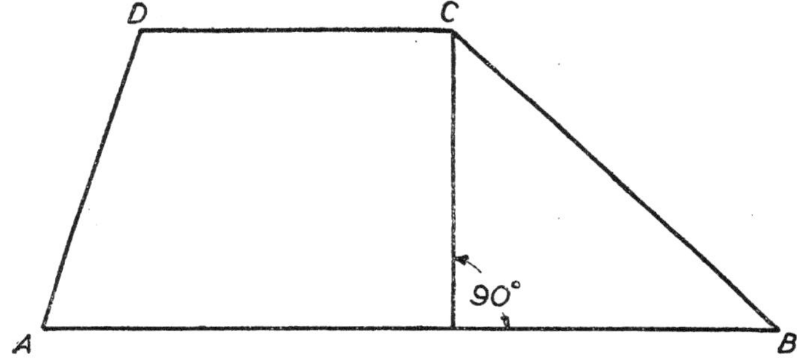

Fig. 78.—Area of trapezium.

To find the area of an *irregular polygon* divide the figure into a number of triangles, calculate the area of each triangle, and add the results together.

Square Root of 2 *and* 3.—As the square root of 3 and the square root of 2 frequently occur in calculations, they should be memorised. The square root of 2 is 1·414, and the square root of 3 is 1·732.

Area of Trapezium.—To find the area of a trapezium divide the figure into two triangles, ABC and ACD (Fig. 78), and calculate the areas in the ordinary way. The altitude of the triangles may be found by erecting a line from C, perpendicular to AB. It will be found after a few examples have been worked out that to obtain the area of a trapezium it is only necessary to *multiply the sum of the parallel sides by one-half the perpendicular distance between them.*

Area of Annulus.—The area of an annulus is the difference of the areas of the two circles. Thus, in Fig. 79, the area of the outer circle is πR^2, and the area of the inner circle is πr^2. Hence :

Area of annulus $=\pi R^2-\pi r^2$.

As π is common to both, we may write :

Area of annulus $=\pi(R^2-r^2)$

$=\pi(R+r)(R-r)$.

Here it will be seen that the bracketed quantity has been

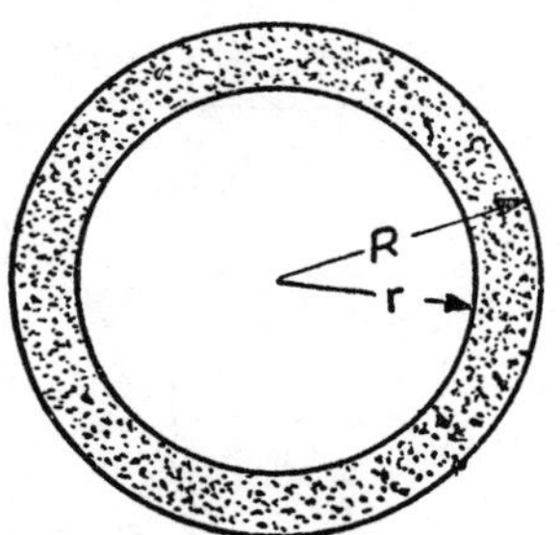

Fig. 79.—Annulus.

factorised, as explained in the chapters on algebra. It will be seen from this last formula that to find the area of an annulus we multiply together the sum and the difference of the two radii and then by π.

Length of Coil of Belting.—The rule to find the length of a rolled coil of belting is $L=\frac{\pi n}{24}(D+d)$, where L=length of belt in feet, $\pi=3\frac{1}{7}$, n=number of turns and fractions of a turn in the coil, D=outside diameter of roll in inches, d=inside diameter of roll in inches. As, however, $\frac{\pi}{24}$ is constant, the formula may be rewritten thus : $L=\cdot 139n\ (D+d)$.

Area of an Ellipse.—The area of an ellipse is πab (Fig. 80). This can be proved in a manner similar to that already given for proving the area of a circle.

Another practical proof is to draw two ellipses of identical size, and to circumscribe one of them with a rectangle. Divide this into four equal rectangles. Now divide one rectangle into seven equal strips, and cut away the shaded part (Fig. 81).

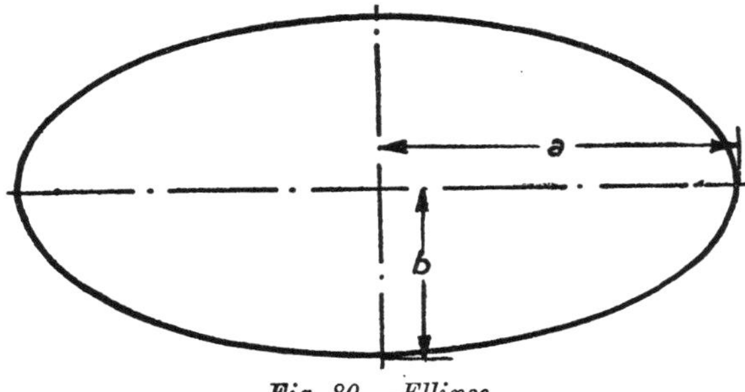

Fig. 80.—Ellipse.

Now cut out the other ellipse, and place it in one pan of a pair of scales, and the other in the second pan of the scales. It will be found that the weights of each are equal. As the shaded

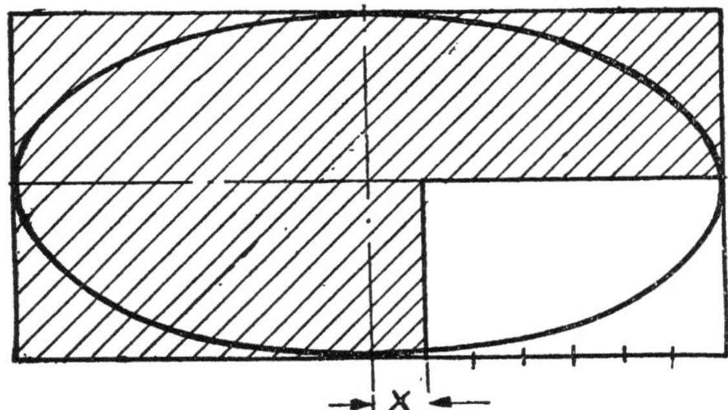

Fig. 81.—Diagram to prove formula for area of ellipse.

portion has an area equal to $3\frac{1}{7}\times$ the area of one rectangle, it is obvious that the area of an ellipse is πab.

The circumference of an ellipse can only be calculated approximately. The formula yielding a fairly close result is :

$$\pi(a+b)$$

where a is half the major axis, and b half the minor axis.

In other words, the length of the circumference of an ellipse is found by multiplying half the sum of the major and minor axis by π.

Centre of Gravity.—The centre of gravity of a body is that point through which the resultant of the gravity of its parts passes in every position a body can assume. It is well known

Fig. 82.—The centre of gravity of a triangle.

that every particle of a given body is attracted by the earth, and the weight of the body represents the total force which gravity exerts on the particles. The forces acting on the body are directed towards the centre of the earth, and their resul-

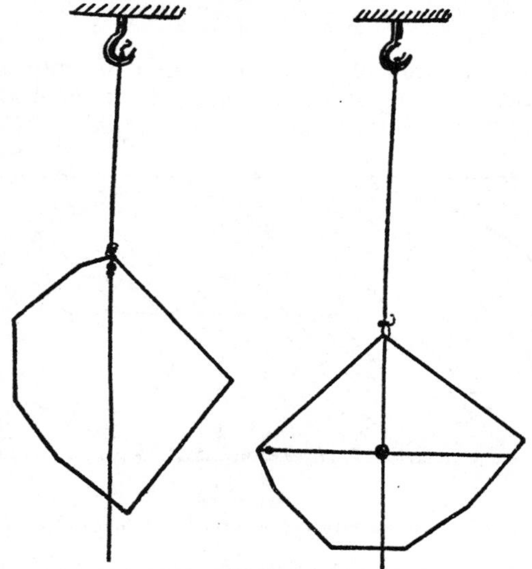

Fig. 83.—Finding centre of gravity of irregular figure.

tant, in other words, the weight of the body, will act through a definite point, known as the centre of gravity. It is obvious that with a symmetrical figure the centre of gravity will coincide with the geometrical centre. With a circular disc of metal, or any other substance, the centre of gravity will be the centre from which the circle is struck, provided the disc is of uniform thickness.

With a triangle, the centre of gravity will be in a line which joins the vertex to the middle point of the opposite side, as in Fig. 82. The centre of gravity of the triangle is the point of intersection of the two bisecting lines, Fig. 82.

With an irregular figure, suspend it from any two points, as in Fig. 83; the point of intersection of the two lines will be the centre of gravity.

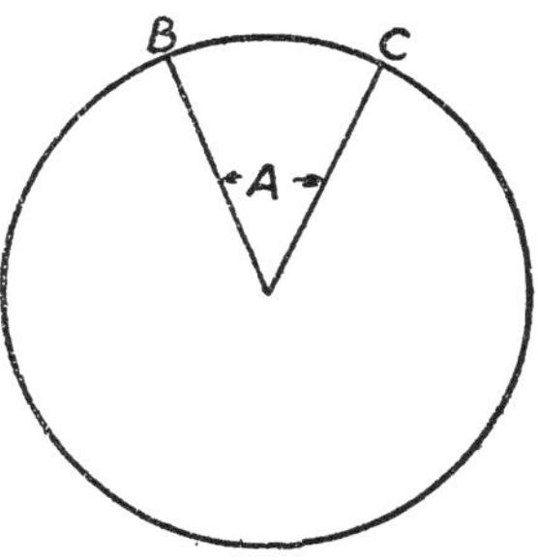

Fig. 84.—Area of segment of circle.

Area of Sector of Circle.—A circle contains 360°. It therefore follows that the area of a sector will bear the same relation to the area of the circle as the *included angle A* bears to 360° (Fig. 84). From this (remembering that the area of a circle is πr^2) it is easy to arrive at the formula for a sector of a circle. The formula is (letting A denote the angle in degrees of the sector):

$$\frac{A}{360}\pi r^2.$$

In circular measure it is more usual to denote the angle by θ. We have seen that $360° = 2\pi$ radians, and therefore we can rewrite the formula:

$$\frac{A}{360}\pi r^2 = \frac{\theta}{2\pi}\pi r^2$$

Example.—The diameter of a circle is 6 ft. What is the area of a sector containing 45° ?

$$\text{Area of circle} = \pi 3^2$$

$$\text{Area of sector} = \frac{45}{360}\pi 3^2$$

$$= \frac{1}{8}\pi 3^2$$

$$= \frac{1}{8} \times 3{\cdot}1416 \times 9$$

$$= 3{\cdot}5343 \text{ sq. ft.}$$

We can also calculate the area of a sector if we are given the length of the arc (such as BC, Fig. 85), for the length of the arc bears the same relation to the circumference as the area of the sector does to the total area. Hence, if the length of the arc BC (Fig. 84) is 4·5 ft. the area of the sector will be

$$\frac{4 \cdot 5}{\pi d} \times \pi r^2$$

$$= \frac{4 \cdot 5}{d} r^2$$

$$= 6 \cdot 75 \text{ sq. ft.}$$

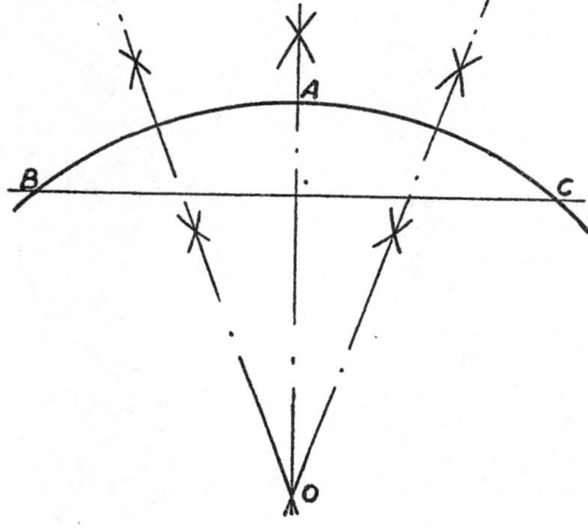

Fig. 85.—Finding radius and length of circular arc.

Length of Circular Arc.—It is often necessary to find the centre of a circle when only a circular arc is given. Let BAC be the arc (Fig. 85). Draw the line BC to cut the arc at any two convenient points. Using B and C as centres and any convenient radius, bisect AC and AB. Produce the bisecting lines, and the point O at which they intersect is the centre of the circle.

When the angle and radius are found the length of the arc can be calculated.

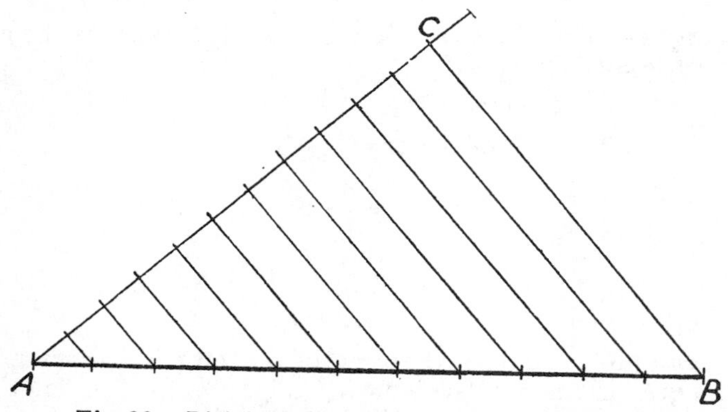

Fig. 86.—Dividing a line into a number of equal parts.

To Divide a Line into Equal Parts.—Suppose it is required to divide a line into a number of equal parts, say 11. Draw a line at any angle from A (Fig. 86) and step off with the dividers (opened to any convenient distance) 11 equal distances, and connect B to C. Lines drawn parallel to BC from the points marked on AC will cut AB into equal divisions.

Finding Radius of Circle from Arc and Chord.—Assume the length of the arc to be 2·5 in., and the chord 2 in.

Let a = length of arc.

c = length of chord.

$$b=1{\cdot}2\,\frac{a-c}{a}$$

Calculate $A=\sqrt{(5b+1{\cdot}25b^2+0{\cdot}625b^3)}$

$$\text{Then radius } r=\frac{a}{2A}$$

In the example given, $a=2{\cdot}5$

$c=2$

$$b=1{\cdot}2\,\frac{2{\cdot}5-2}{2{\cdot}5}=0{\cdot}24$$

$$A=\sqrt{(1{\cdot}2+1{\cdot}25\times0{\cdot}24^2+0{\cdot}625\times0{\cdot}24^2)}$$
$$=\sqrt{(1{\cdot}2+0{\cdot}072+0{\cdot}00865)}$$
$$=\sqrt{1{\cdot}28}=1{\cdot}131$$

$$\text{and so } r=\frac{2{\cdot}5}{2\times1{\cdot}131}=1{\cdot}103$$

Calculating Altitude of Triangle from Two Angles greater than 90°.—A trigonometrical problem often encountered is illustrated in Fig. 86a, where it is necessary to calculate the altitude from two angles greater than 90°. The following is the method of solving such problems :

$DB=AB\cot 32$

$CB=AB\cot 42$

$CD=DB-CB=AB\cot 32-AB\cot 42$

$=AB\,(\cot 32-\cot 42)$.

Assuming DC in Fig. 86A to be 200 ft. :

$200=AB\,(1{\cdot}6003-1{\cdot}1106)$

$$\therefore AB=\frac{200}{1{\cdot}6003-1{\cdot}1106}=\frac{200}{{\cdot}4897}=408 \text{ ft.}$$

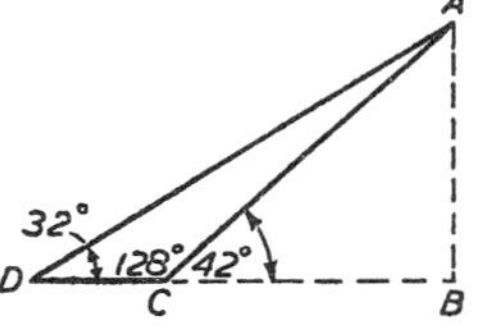

Fig. 86a.—Calculating height from two angles greater than 90°.

CHAPTER XXV

VOLUME—WEIGHT—MASS—DENSITY—SOLIDS

Volume, Weight, and Density.—It is well known that *the volume of a rectangular solid is found by multiplying the length, breadth, and height together*; the units of length, breadth, and height must, of course, be similar. If one dimension is given in ft. and another in in. the ft. must be reduced to in. or the in. to ft. Thus, in Fig. 87, which represents 1 cubic ft., the *volume or cubical content* is $12 \times 12 \times 12 = 1728$ cubic in.

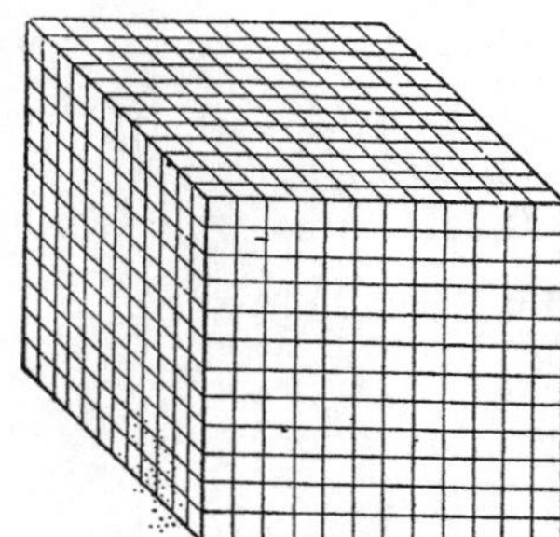

Fig. 87.—Volume of cube.

The *gallon* is the British unit of *volume and capacity*; it is the volume occupied by 10 lb. of chemically pure water at a temperature of 62° F. Nearly all bodies expand when hot and contract when cold.

Thus, when hot, 1 cubic ft. of water would occupy a greater volume than 1 cubic ft., and when cooled to a temperature lower than 62° F. it would occupy less than 1 cubic ft. Hence, it is necessary to state the temperature of the water in defining its volume.

At 32° F. the weight of a cubic ft. of water is 62·418 lb.

At 62° F. the weight of a cubic ft. of water is 62·355 lb.

At 212° F. the weight of a cubic ft. of water is 59·64 lb.

For ordinary calculations the weight may be taken as 62·3 lb. Often the weight of 1 cubic ft. of water is taken as 1000 oz. (62·5 lb.), which equals $6\frac{1}{4}$ gallons.

A pint of water weighs $1\frac{1}{4}$ lb. approx.

The metric unit of volume is the *litre*, which is (very approximately) a cubic decimetre (see section on the metric system), or 1000 cubic centimetres. This equals 1·76 English pints.

Mass.—The quantity of matter a body contains is known as its *mass*. The *pound avoirdupois* is the British unit of mass.

The metric unit of mass is the *kilogram*. It is the mass of a platinum cylinder deposited in the French archives.

A litre of pure water weighs, at 4° C., 1 kilogram, or 1000 grams.

The *weight* of a body is defined as the *attractive force* exerted (at the surface of the earth) upon it. The force is greatest at the poles and least at the equator. This force is known as *gravity*. The force of gravity varies with the distance from the centre of the earth. At this point the force of gravity is nil, and there the weight of the body would also be nil.

The acceleration due to g, of a free falling body, is 32·2 ft. per sec./per sec.

Density.—The mass of the unit of volume of a body is known as its *density*. If the unit of mass is 1 lb., and the unit of volume 1 cu. ft., then the density is the number of pounds in a cubic foot of the substance ; in the metric system the density is the number of grams in a cubic centimetre of the substance.

Specific Gravity.—The *relative density* of a substance is *the ratio of its weight to that of an equal volume of another, and standard, substance*. Distilled water is the standard substance adopted. The relative density of a substance is known as its *specific gravity*. If the specific gravity of a substance is stated to be 9·7, then the weight of 1 cu. ft. of that substance is 9·7 times the weight of a cu. ft. of pure water. Hence the weight would be :

$$9{\cdot}7 \times 62{\cdot}3 = 604{\cdot}31 \text{ lb.}$$

Volumes of Solids.—The *regular solids* are the *cube*, *tetrahedron*, *octahedron*, *dodecahedron*, and *icosahedron*.

A cube has six equal and square faces.

A tetrahedron has four equal faces, each of which is an equilateral triangle.

The octahedron has eight equal faces, each of which is an equilateral triangle.

The dodecahedron has twelve equal faces, each of which is a pentagon.

The icosahedron has twenty equal faces, each of which is an equilateral triangle.

Other solids are the *cone*, *cylinder* *pyramid*, *sphere*, and *prism* or *parallelopiped*.

When one end of a line, as XY (Fig. 88), passes through a fixed point, whilst the other describes a curve, the figure described is a pyramid ; and if the curve described is a circle,

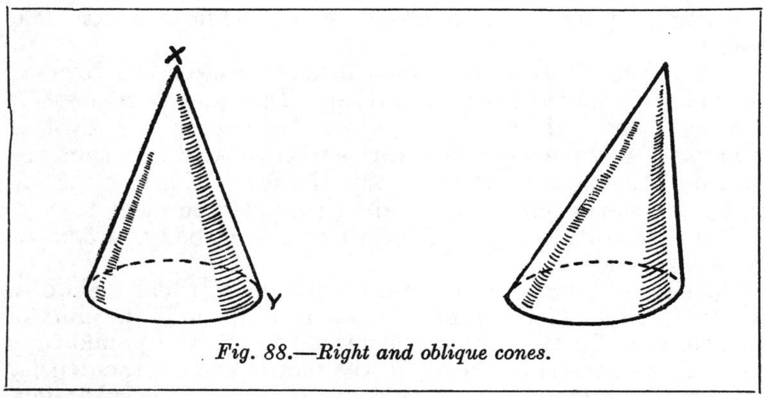

Fig. 88.—Right and oblique cones.

the figure traced will be a *right cone* if the fixed point is exactly over the centre of the circle. If the fixed point is not so located, the figure traced will be an *oblique cone.*

Various prisms are shown in Figs. 89 to 92.

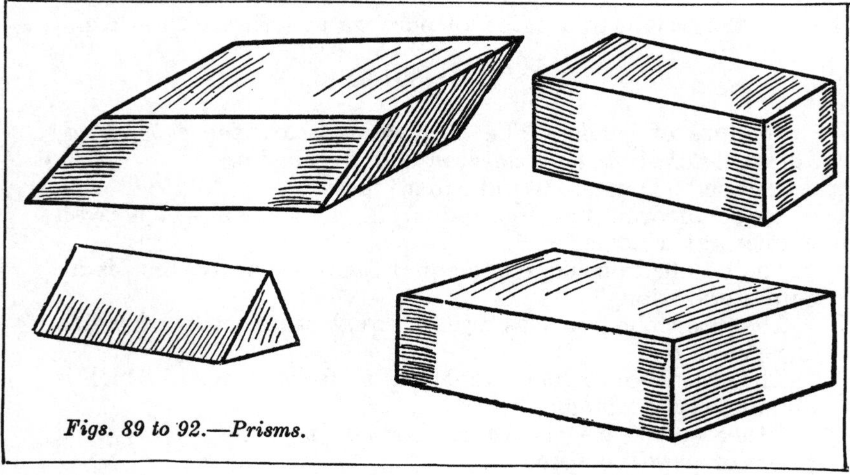

Figs. 89 to 92.—Prisms.

If a line, as XY (Fig. 93), traces the contour of any *rectilinear polygon*, the figure so described is a prism and the prism is designated according to the shape of its ends. Thus, there are hexagonal, pentagonal, triangular, etc., prisms.

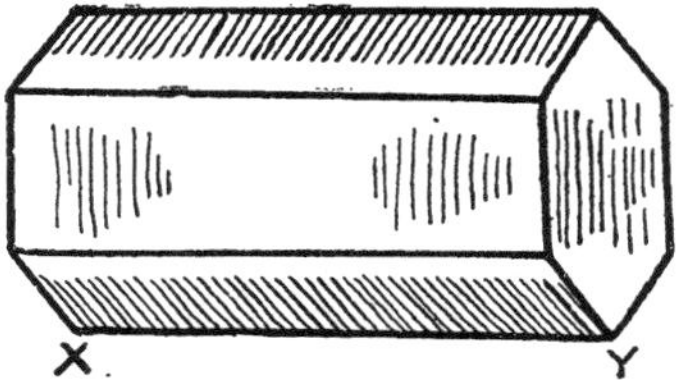

Fig. 93.—Regular polygon.

A *right prism* (sometimes termed a *parallelopiped*) has its ends perpendicular to its side faces. If the ends are not perpendicular it is termed an *oblique prism*.

The volume of a right prism is found by multiplying together the length, breadth, and height ; and the volume of an oblique prism is the area of the base multiplied by the altitude (Fig. 94).

Fig. 94.—Oblique prism.

The volume of a cylinder is found by *multiplying the area of the base by the height*, or :

$$\pi r^2 h$$

The surface of a cylinder consists of the *area of the two ends, plus the area of the curved surface.*

Area of curved surface	$=\pi dh$, or $2\pi rh$
Area of two ends	$=2\pi r^2$
Total area	$=2\pi rh+2\pi r^2$
	$=2\pi r(h+r)$

The volume of an *oblique cylinder* (Fig. 95) is :

$$\text{Area of base} \times \text{height} = \pi r^2 h$$

A cylinder cut by a plane not parallel to the base is termed a *frustum of a cylinder.* The volume of a frustum of a cylinder

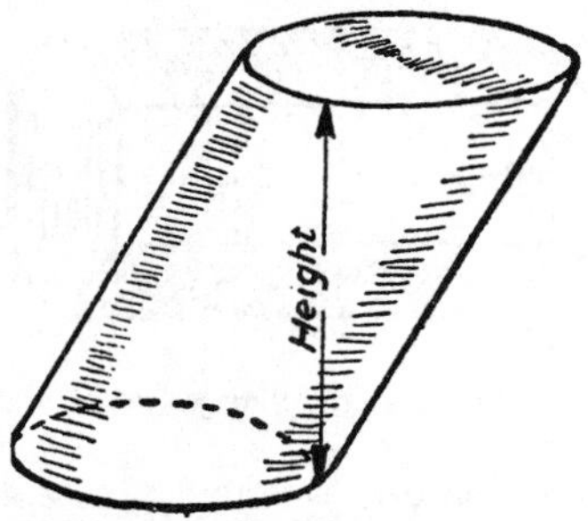

Fig. 95.—Oblique cylinder.

(Figs. 96 and 97) is found by multiplying the area of the base by the mean height h.

$$\text{Volume of frustum of cylinder} = \pi r^2 h = \pi r^2 \left(\frac{a+b}{2}\right)$$

The end surface of a cylinder cut by a plane not parallel to the base is an ellipse.

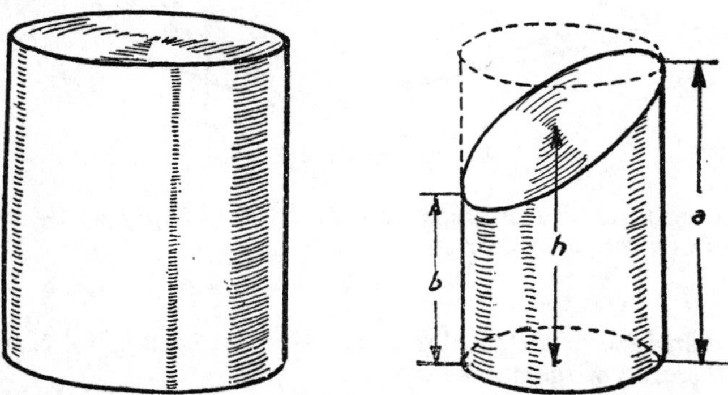

Figs. 96 and 97.—Cylinder and frustum of cylinder.

The volume of a hollow cylinder is obviously the volume of the outer cylinder, less the volume of the imaginary inner cylinder$=\cdot 7854\ (D^2-d^2)h$.

Pyramids.—Draw a cube as Fig. 98, and from a point representing the centre of the cube construct the square pyramid shown. From inspection of the diagram it is obvious that the cube contains six such pyramids, and therefore the volume of the pyramid is $\frac{1}{6}$ the volume of the cube.

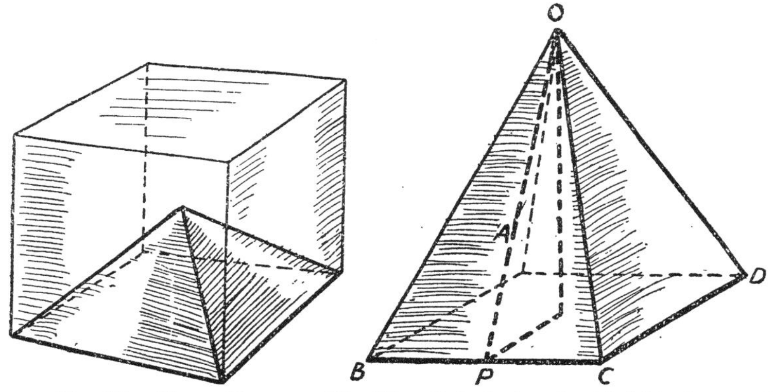

Fig. 98.—Pyramid. *Fig. 99.—Calculating slant height of pyramid*

If the length of one side of the cube be x, then the volume of the pyramid will be :

$$\frac{1}{6}x^3$$

Taking half the cube, the volume of the pyramid will be :

$$\frac{1}{3}x^2 \times \frac{x}{2} = \frac{1}{3}x^2h$$

(h being the height of the pyramid).
Therefore,

$$\text{Volume of pyramid} = \tfrac{1}{3}(\text{area of base} \times \text{height}).$$

The total surface area of any pyramid is the sum of the areas of the triangles plus the area of the base, and if the latter is a regular polygon, the triangles will be isosceles triangles. Draw perpendiculars from the *vertex O* (Fig. 99), from which the area of each triangle will be :

$$\frac{\text{Length of side} \times OP}{2}$$

The total surface of all the faces will be :

$$\left(\frac{\text{sum of all the sides}}{2}\right)OP$$

Letting l equal OP, and a the length of one side,

$$\text{Total area of all triangles}=\frac{l}{2}\Sigma a$$

Let n represent the number of sides,

$$\text{Then lateral surface of all the triangles}=\frac{nal}{2}$$

In other words, *the total surface of the triangles equals half the perimeter of the base* × *the slant height.*

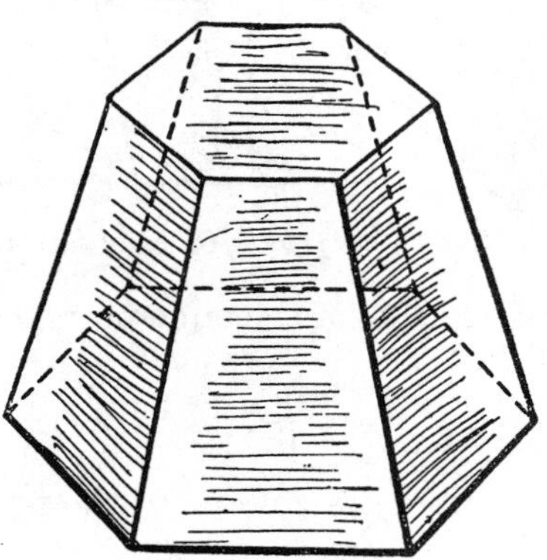

Fig. 100.—Pyramid.

The total area = area of base + area of triangles.

The *slant height OP* (Fig. 99) can be calculated from the trigonometrical formula previously given.

If a pyramid be cut across a plane parallel to its axis, each face will form a trapezium (Fig. 100). We have already seen that the area of a trapezium is found by multiplying half the sum of the parallel sides by the distance separating them. Let this distance be l, the length of the smaller side ab, and the longer AB. Then :

$$\text{Area of each trapezium}=\frac{ab+AB}{2}\times l$$

And, letting n represent the number of sides, the total surface of all the trapeziums will be :

$$\frac{n}{2}(ab+AB)l$$

Expressed as a rule, *the total area of the trapeziums is found by multiplying the sum of the perimeters of the end polygons by half the distance between them.*

Cone.—A pyramid on a circular base is a cone and the same rule for volume applies :

$$\text{Vol. of cone}=\frac{1}{3}(\text{area of base and height})$$

$$=\frac{1}{3}\pi r^2 h$$

The volume of a cone is therefore one-third the volume of a cylinder of the same base and height.

The lateral surface of a cone is found by multiplying the circumference of the base by the slant height and dividing by 2.

The slant height of a cone is found from the formula :

$$s=\sqrt{(h^2+r^2)}$$

where s=slant height, h=vertical height of cone, and r= radius of base.

As with a frustum of a pyramid, the curved surface of a cone may be considered as consisting of a number of trapeziums, the circular ends forming the parallel sides, and the slant side being the distance between them. Therefore, the curved surface of a frustum of a cone

$$=\tfrac{1}{2}(\text{sum of perimeters of ends}\times\text{slant side})$$
$$=\tfrac{1}{2}(2\pi r+2\pi R)l$$
$$=\pi(R+r)l$$

R and r represent the radius of the base and of the small end respectively.

Total area (curve surface plus area of ends) is found from the formula :

$$\pi(R+r)l+\pi(R^2+r^2)$$

Solid Ring or Torus.—A solid ring may be considered as a cylinder bent into circular form, and it is apparent (Fig. 101) that the height of such a cylinder will be equal to the mean diameter of the ring. The mean diameter may be found by adding together the inner and outer diameters and dividing by two.

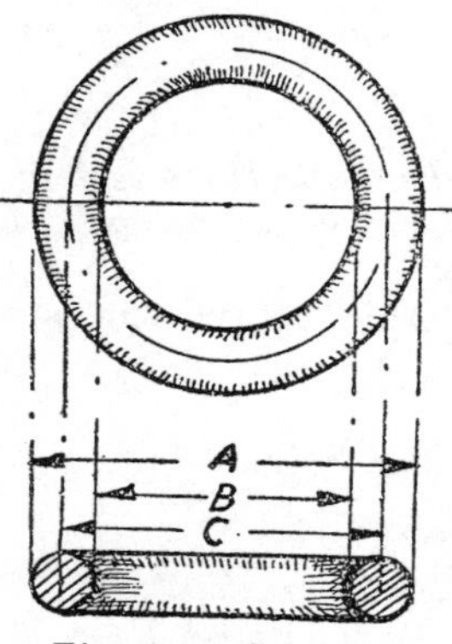

Fig. 101.—Solid ring.

The surface area is equal to the circumference of a cross section multiplied by the mean length.

The mean length is πC.

Mean diameter is $\dfrac{A+B}{2}$.

Therefore, total area$=2\pi r\times 2\pi R$, where r=radius of cross section, and R=mean radius.

$$2\pi r\times 2\pi R=4\pi^2 Rr$$

The volume of a solid ring is the area of a cross section multiplied by the mean length.

Area of cross section$=\pi r^2$.
Mean length $=2\pi R$.
Volume $=\pi r^2\times 2\pi R$
$=2\pi^2 Rr^2$.

Also, if the solid ring is rectangular the mean diameter is multiplied by the area of the cross section to obtain the volume.

Sphere.—A sphere is a *semi-circle* rotated about its axis (the diameter). Any plane section of a sphere is a circle, and if the plane passes through the centre of a sphere, the section will be a *great circle.* Now, the area of the great circle is, as we have seen, πr^2, and the area of a *hemisphere*

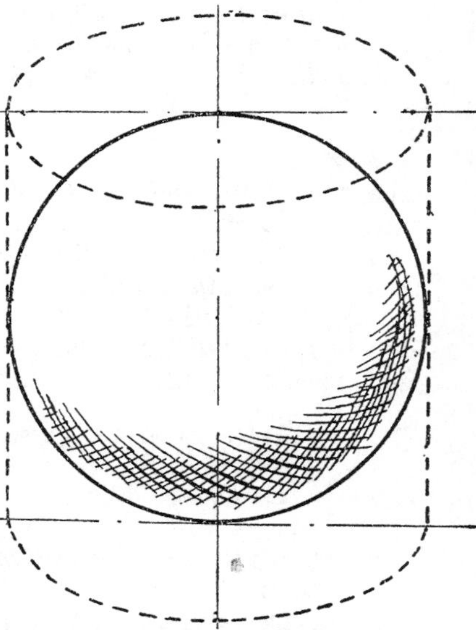
Fig. 102.—The surface area of a sphere is equal to the surface area of its circumscribing cylinder.

is $2\pi r^2$. Therefore, the area of a sphere will be twice that, namely $4\pi r^2$, or πd^2.

The area of a sphere is the same as the area of a cylinder which exactly encloses the *sphere* (Fig. 102).

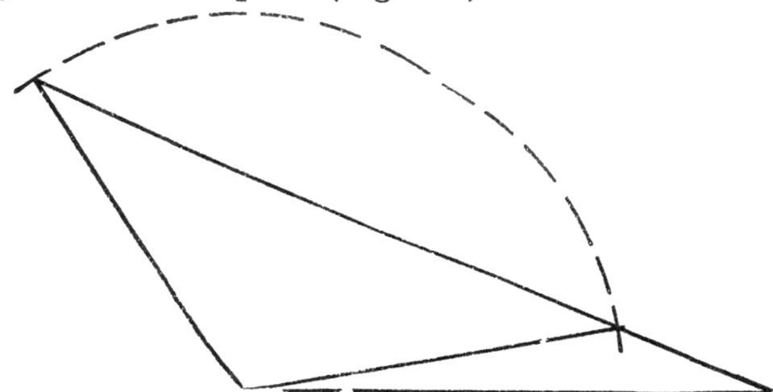

Fig. 103.—The ambiguous case, when two sides of a triangle and one angle are given ; there can be two solutions, as shown here.

The volume of a sphere is $\frac{2}{3}$ the volume of the cylinder enclosing it.

The volume of a cylinder is $2\pi r^3$, and dividing this by $\frac{2}{3}r$, we obtain $\frac{4}{3}\pi r^3$, which equals $\frac{\pi}{6}d^3$.

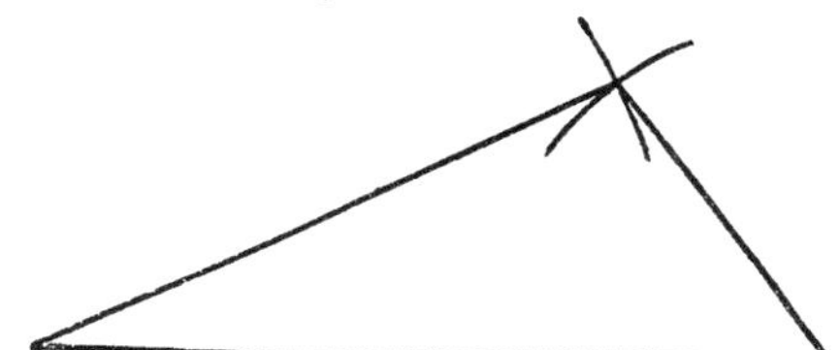

Fig. 104.—To construct a triangle where the lengths of the three sides are known, draw one side, and from each end describe arcs of a radius equal to the length of each of the other sides.

Circumscribing Circle.—The radius of a circle which may be inscribed in a triangle is found from the formula :

$$R = \frac{abc}{4 \times \text{area of triangle}}$$

where a, b, and c are the lengths of the sides of the triangle.

$$2R = \frac{a}{\sin A} = \frac{b}{\sin B} = \frac{c}{\sin C}$$

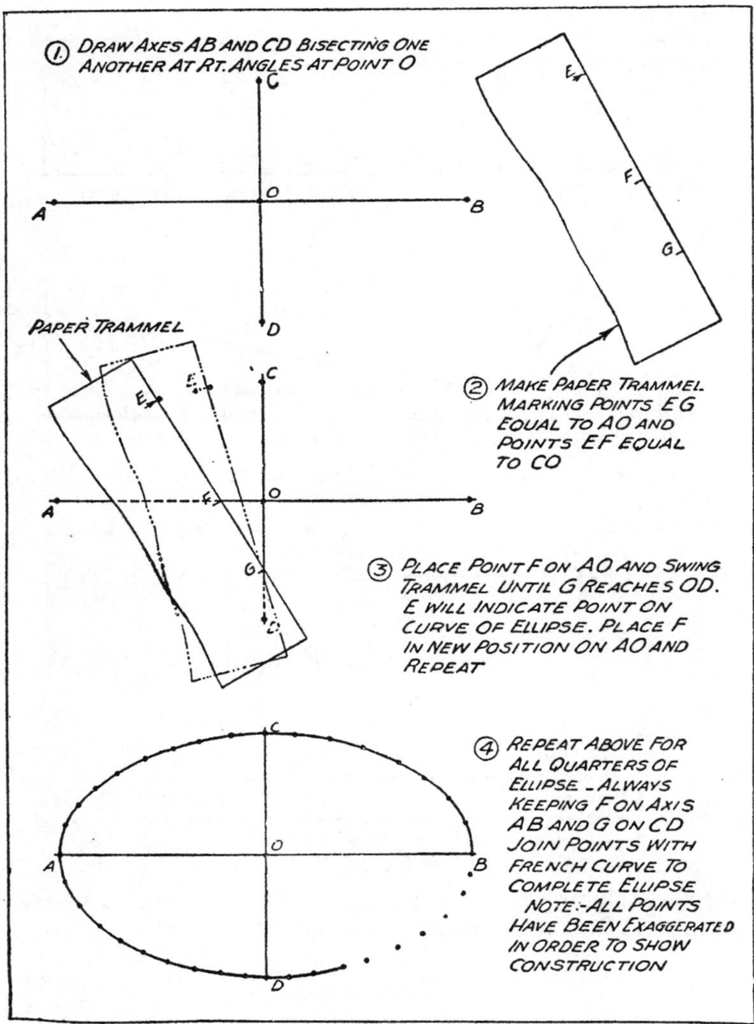

① DRAW AXES AB AND CD BISECTING ONE ANOTHER AT RT. ANGLES AT POINT O
C
O
A
B
D
E
F
G
PAPER TRAMMEL
② MAKE PAPER TRAMMEL MARKING POINTS EG EQUAL TO AO AND POINTS EF EQUAL TO CO
③ PLACE POINT F ON AO AND SWING TRAMMEL UNTIL G REACHES OD. E WILL INDICATE POINT ON CURVE OF ELLIPSE. PLACE F IN NEW POSITION ON AO AND REPEAT
④ REPEAT ABOVE FOR ALL QUARTERS OF ELLIPSE - ALWAYS KEEPING F ON AXIS AB AND G ON CD JOIN POINTS WITH FRENCH CURVE TO COMPLETE ELLIPSE
NOTE:-ALL POINTS HAVE BEEN EXAGGERATED IN ORDER TO SHOW CONSTRUCTION

Figs. 105 to 108.—One method of drawing an ellipse.

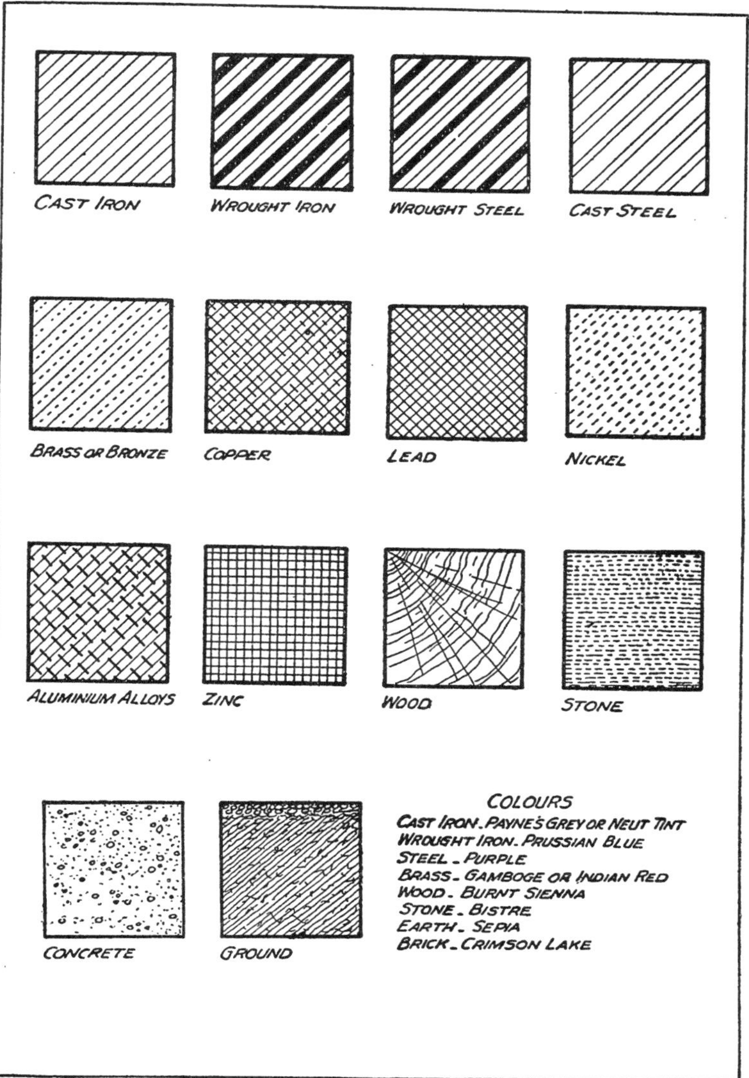

Figs. 109 to 122.—The conventional methods of cross-hatching various materials in sectional drawings.

Inscribed Circle.—This is found from the formula :

$$R=\frac{\text{Area of triangle}}{s}$$

$$=\sqrt{\frac{(s-a)(s-b)(s-c)}{s}}$$

$$=(s-a)\tan\frac{A}{2}=(s-b)\tan\frac{B}{2}=(s-c)\tan\frac{C}{2}$$

where s=sum of lengths of sides.

Escribed Circle.—To find the radius of the circle touching side a and sides b and c produced :

$$\frac{\text{area of triangle}}{s-a}=\sqrt{\frac{s(s-b)(s-c)}{(s-a)}}$$

Figs. 103 and 104 show two examples of the construction of triangles. It is very important to remember the point illustrated in Fig. 104, as it frequently occurs in examinations.

Area of Irregular Figures.—A reliable method of calculating the area of irregular figures, such as the diagram $ABDC$, is given on page 167.

Divide the base line AB into any convenient number of equal parts. The greater the number the more accurate the result. At the centre of each part erect ordinates, as shown dotted. Now measure the height of the ordinates *aa*, *bb*, *cc*, *dd*, *ee*, *ff*, *gg*, *hh*, and add them together. Divide their sum by the number of ordinates. This will give the mean height H, or mean ordinate AJ. A simple method of obtaining the total length of the ordinates is to use the edge of a piece of paper, starting with *aa* and adding *bb*, *cc*, etc. The area is then found by multiplying AB by AJ.

The Trapezoïdal Method.—Divide the base into a number of equal parts (say 8), spaced a distance s apart. Let the height of each ordinate be h_1, h_2, h_3, h_4, etc., and the mean height h. Then :

$$h=\frac{s}{8}\left\{\tfrac{1}{2}(h_1+h_9)+h_2+h_3+h_4+h_5+h_6+h_7+h_8\right\}$$

(The number of ordinates is always 1 more than the number of parts.) Expressed as a rule : *Divide the base into any number of equal parts, and add half the sum of the end ordinates to the sum of all the others. Multiply the result by the common interval s to obtain the area ; or divide the result by the number of spaces to obtain the mean ordinate.*

Mean ordinate × length = area of figure.

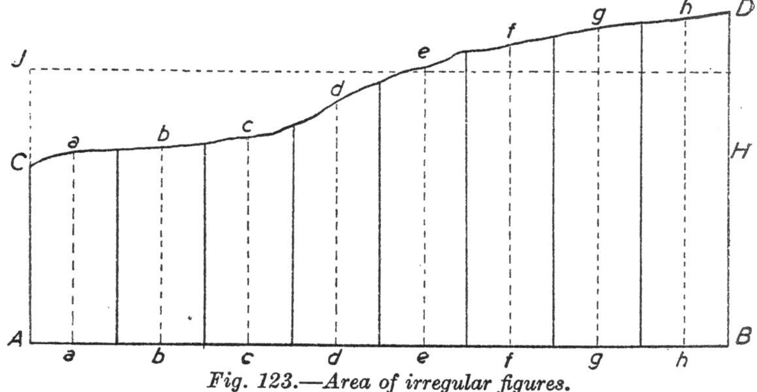

Fig. 123.—Area of irregular figures.

Simpson's Rule.—Probably the most accurate method. Divide the base AB into an even number of equal parts (say, 6) to produce an odd number of ordinates (7), spaced a distance s apart. Then :

Area $ABDJ=\frac{s}{3}\left\{h_1+h_7+4(h_2+h_4+h_6)+2(h_3+h^2{}_5)\right\}$

This reduces to :

$$\frac{s}{3}(A+4B+2C)$$

where A=sum of first and last ordinates,
B=sum of even ordinates,
C=sum of odd ordinates.

Expressed as a rule :

Add together the extreme ordinates, four times the sum of the even ordinates, and twice the sum of the odd ordinates (omitting the first and last) ; then multiply the result by one-third the space between the ordinates.

In the case of an irregular figure in which the end ordinates are zero, then A is zero, and the formula then becomes :

$$\frac{s}{3}(0+4B+2C)$$

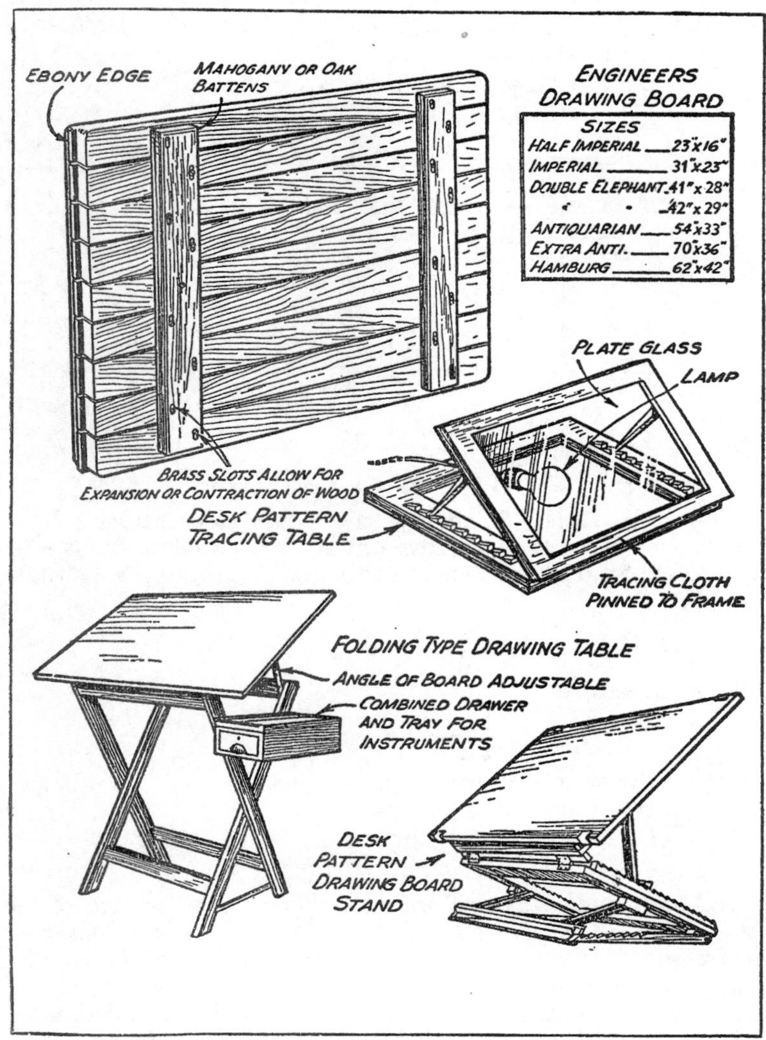

Figs. 124 to 128.—Drawing-boards, tables, etc.

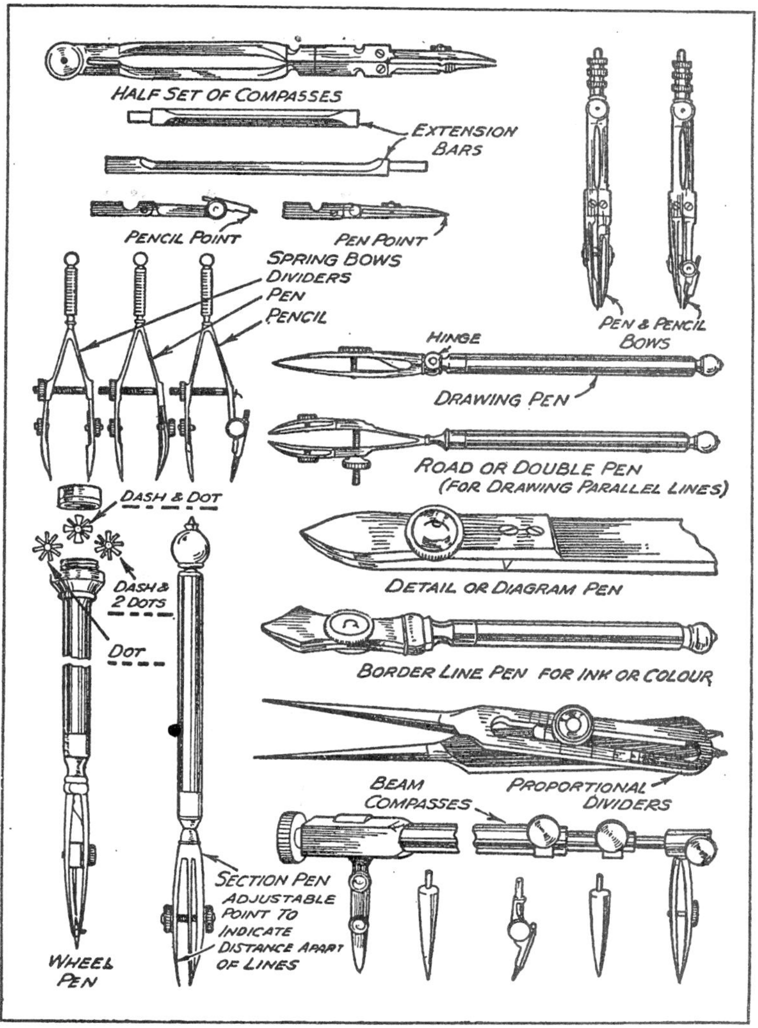

Figs. 129 to 145.—Various drawing instruments.

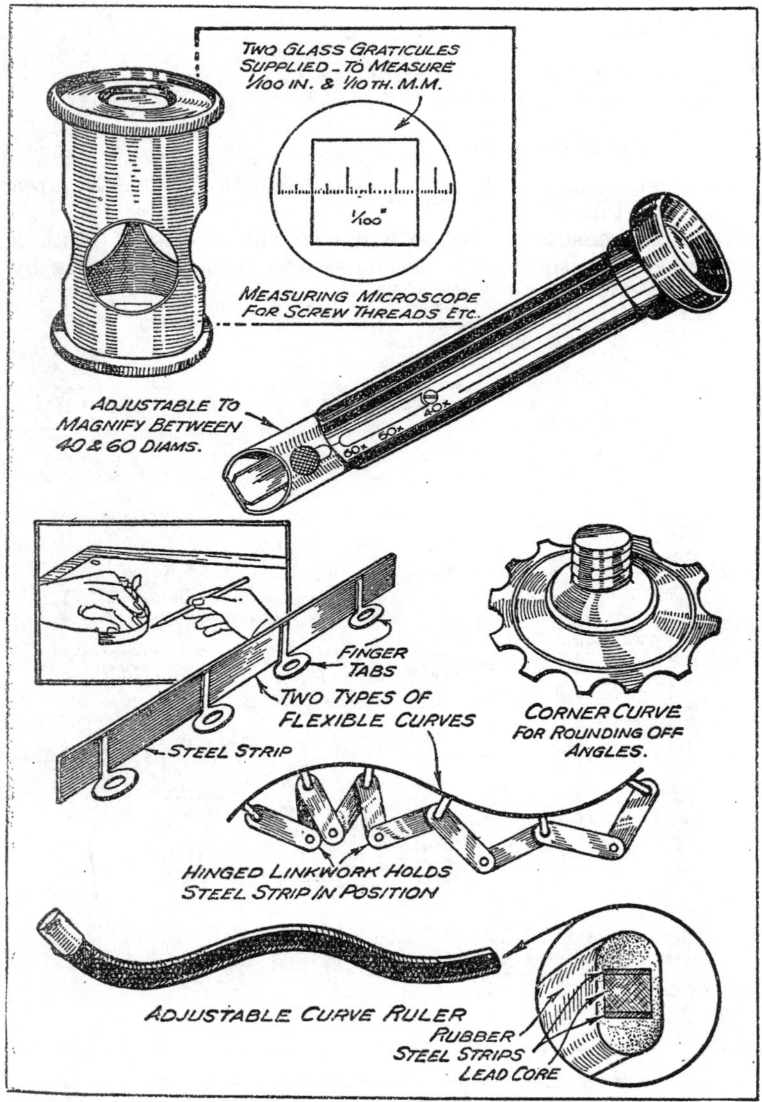

Figs. 146 to 153.—Various useful devices for drawing, etc.

In some cases, where the irregular figure is bounded by two curved lines, it is convenient to divide the figure into two parts, and calculate the area of each independently.

When the area is completely bounded by an irregular curve, parallel lines are drawn, touching the top and bottom of the curve, and vertical lines touching the sides. The figure is thus enclosed in a rectangle, and by means of ordinates the area can be found as before.

For any reasonably smooth curve the ordinate y at a distance x from the origin can be represented very nearly by an expression of the form $y=a+bx+cx^2$.

Now the area of any vertical strip of height y and width dx is ydx and so the area under the curve between $x=0$ and $x=L$ is—

$$\int_0^L ydx=\int_0^L (a+bx+cx^2)dx=aL+\frac{bL^2}{2}+\frac{cL^3}{3}$$

$$=\frac{L}{6}(6a+3bL+2cL^2).$$

Now the ordinate at the left-hand end where $x=0$ is $a+b\times 0+c\times 0^2=a$.

The ordinate at the right-hand end is $a+bL+cL^2$.

The sum of these is $2a+bL+cL^2$, and subtracting this from the quantity in the round bracket above the remainder is $4a+2bL+cL^2=4\left(a+\frac{bL}{2}+\frac{cL^2}{4}\right)$.

The mid-ordinate is $a+b\left(\frac{L}{2}\right)+c\left(\frac{L^2}{2}\right)$ which is just one-quarter of the remainder. Hence

$$\text{Area}=\frac{L}{6}\begin{pmatrix}\text{left-hand ordinate}+\\ 4\times\text{mid-ordinate}+\\ \text{right-hand ordinate}\end{pmatrix}$$

If a figure is divided into four equal widths the formula is applied first to the left-hand half and then to the right-hand half, and the total area for an overall length L is then—

$$\frac{1}{6}\frac{L}{2}(A+4B+C)+\frac{1}{2}\frac{L}{6}(C+4D+E)$$

$$=\frac{L}{12}(A+4B+2C+4D+E).$$

Where A and E are the end ordinates, and B, C, and D the equidistant intermediate ones.

CHAPTER XXVI

The Infinitesimal Calculus : Differentiation

Technical students who, in the course of their mathematical studies, have made some acquaintance with the calculus and find it a little puzzling at first, are often disposed to ask, " What is the use of it ? " This question is not easy to answer in a few words, but it may clear the ground to say at the outset that in very few calculations in engineering design is it necessary to use the " infinitesimal calculus "—to give it its full name. On the other hand, a working knowledge of the calculus may occasionally save time in calculation. The position is something the same as if a child were to ask what is the use of the multiplication tables. It might be said in reply that the multiplication tables are not *essential* (because what they do can also be done by somewhat laborious adding), but they are extremely useful in saving time.

Generally speaking, use of the calculus in mathematical work is advantageous, not because it gives a result that cannot be got in any other way, but because it gives the result with less labour (and often with higher accuracy) than is possible by more elementary methods. The designer need rarely be beaten because he knows no calculus, but he can sometimes do things more quickly if he can use it. Here again it may be interjected that a theoretical knowledge of the calculus is not quite the same thing as ability to use it advantageously in practical calculations.

The subject of the infinitesimal calculus is divisible into two main subjects, " differential calculus " and " integral calculus," which are related to each other in much the same way as are multiplication and division in arithmetic. It may be noted that the word " calculus " literally implies nothing more than a system of calculation, but its present-day use implies " calculus of infinitesimals," and the meaning of that is " mathematical relations between infinitely small quantities." It is here, right at the very beginning of the subject, that students' doubts may arise because they realise that infinitely small quantities have no essential place in practical engineering. The objection is not a serious one, however. It has a parallel in the subject of mechanics, where attention is often devoted to movements of "particles" connected by "weightless inextensible strings" and sliding over "perfectly smooth surfaces." In each case these unpractical subjects are selected for discussion so that the student may concentrate at first on fundamentals, leaving the entanglements of practical details to be overcome later.

Scope of the Differential Calculus.—The calculus is used in dealing with quantities that change or that may be changed. In engineering practice there are many quantities that change continuously; for example, the pressure of the gas in the cylinder of an internal combustion engine. In engineering design everything is likely to change because the basic procedure is very often "cut and try." Certain dimensions are assumed (i.e. guessed), the main elements of the design are based on them, and the result reviewed. If it is not satisfactory in some respect, changes are made in one or more of the assumed quantities and the results of such changes are investigated. It is necessary to answer the question, If we change A by so much, how much will B and C be altered? If there are known mathematical relations between A, B, and C, the differential calculus will answer this question more quickly than will the obvious way of going through all the original calculations again on the basis of the new value of A. In brief, the change produced in B by a certain change in A is equal to that change multiplied by the "differential coefficient of B with respect to A." This quantity with the long name, "differential coefficient of B with respect to A," is denoted in mathematical shorthand by $\frac{dB}{dA}$, and an essential part of the subject of the differential calculus is concerned with the rules for determining the differential coefficients of mathematical functions of a quantity.

Function.—The word "function" used in the last sentence has a special and clearly definable meaning. If the value of a quantity B can be expressed mathematically in terms of a quantity A—that is to say, if we have a mathematical expression whose value is always equal to that of B when the appropriate value of A is inserted in it—then B is said to be a "function of A." It is equally true to say that A is a function of B.

Rate of Change.—When the value of A is changed, the value of B will (nearly always) change, and the change in B divided by the change in A is the "rate of change of B with respect to A," and it is the differential coefficient of B with respect to A. By mathematical convention, dB means (not in this case d times B) "change in B," and dA is the corresponding change in A.

Now, strictly speaking, rate of change is the same thing as "ratio of changes" only when those changes are very small ones, and in fact, unless we specify that the changes shall be infinitely small, the ratio of them has no definite value.

For example, let a quantity y be connected with a quantity x by the relation $y=5x^2$. Suppose that in the first instance $x=2$, then y is equal to 20.

Now consider what happens when x undergoes certain typical changes as worked out below in tabular form.

Original $x=2$. Original $y=20$.

Change in x	New x	New y	Change in y	$\frac{\text{Change in } y}{\text{Change in } x}$
1	3	45	25	25
0·5	2·5	31·25	11·25	22·5
0·25	2·25	25·313	5·313	21·25
0·1	2·1	22·05	2·05	20·5
0·01	2·01	20·2005	0·2005	20·05

Here are five changes in x ranging from 1 down to 0·01. It will be noticed that the value of the ratio (change in y) (change in x) is not constant, but varies with the amount of the change. As the change in x diminishes towards zero, the change in the ratio also diminishes, not towards zero but (apparently) towards the value 20. Actually it can be proved mathematically that the ratio does become 20 when the change in x is infinitely small. It is necessary to use some such expression as this last one and not to say "zero," because otherwise it could be objected that if the change is zero there is no change at all, and discussion of ratios of changes becomes meaningless.

So the differential coefficient properly so called is the ratio of the changes when they are infinitely small. A good deal of the usefulness of differential calculus in practical mathematics is, however, dependent on the fact that the ratio of the changes is nearly equal to the differential coefficient even when the changes are not infinitely small. For example, the table shows that when x changes by 0·1, y changes by as much as 2·05, which is certainly not infinitely small, but the ratio of the changes is 20·5, which is nearly equal to the differential coefficient, 20.

Determination of Differential Coefficient.—The numerical example discussed above shows how the value of the differential coefficient can be determined (very nearly) in any particular instance by purely arithmetical work. Can it be determined for any given mathematical function by a straightforward mathematical process ? The answer is "Yes," the process is called "differentiation," and it can always be carried out by following a few simple rules, no matter how complicated the mathematical function may be.

In the following, the variable quantities are x and y; the others are constant :

$$y=(ax+b)^n \qquad \frac{dy}{dx}=na(ax+b)^{n-1} \qquad (1)$$

$$y=a\sin(nx+b) \qquad \frac{dy}{dx}=na\cos(nx+b) \qquad (2)$$

$$y=a\cos(nx+b) \qquad \frac{dy}{dx}=-na\sin(nx+b) \qquad (3)$$

$$y=a\tan(nx+b) \qquad \frac{dy}{dx}=na\sec^2(nx+b) \qquad (4)$$

$$y=ab^{(cx+d)} \qquad \frac{dy}{dx}=acb^{(cx+d)}\log_e b \qquad (5)$$

The quantity e in (5) is the "base of Napierian logarithms," 2·71828.

These formulæ cover many of the cases encountered in engineering mathematics. For example, if $y=4x^3$ the differential coefficient of y with respect to x is determined by the use of (1), the values of the constants in this special case being $a=4$, $b=0$, $n=3$, and so

$$\frac{dy}{dx}=12(4x+0)^2=192x$$

Again, if $y=12 \cos(-2x+7)$, the differential coefficient is determined by the use of (3), the values of the constants being $a=12$, $n=-2$, $b=7$, and so

$$\frac{dy}{dx}=24 \sin(-2x+7)$$

In addition to the formulæ given above, there are certain rules that make it possible to differentiate combinations of functions. In the following, u, v, w, etc., are functions of x :

$$y=u+v+w \ldots \text{etc.} \qquad \frac{dy}{dx}=\frac{du}{dx}+\frac{dv}{dx}+\frac{dw}{dx}+\ldots \text{etc.} \qquad (6)$$

$$y=uvw \ldots \text{etc.}$$

$$\frac{dy}{dx}=(uvw \ldots \text{etc.})\left[\frac{1}{u}\frac{du}{dx}+\frac{1}{v}\frac{dv}{dx}+\frac{1}{w}\frac{dw}{dx}+\ldots \text{etc.}\right] \qquad (7)$$

$$y=u/v \qquad \frac{dy}{dx}=\left(v\frac{du}{dx}-u\frac{dv}{dx}\right)/v^2 \qquad (8)$$

For example, if $y=(x+1) \sin(2x-1)$, this is dealt with by means of (7), giving

$$\frac{dy}{dx}=(x+1)\sin(2x-1)\left[\frac{1}{x+1}\frac{d(x+1)}{dx}+\frac{1}{\sin(2x-1)}\frac{d\sin(2x-1)}{dx}\right]$$

Using (1) and (2) this becomes—

$$\frac{dy}{dx}=(x+1)\sin(2x-1)\left[\frac{1}{x+1}+\frac{2\cos(2x-1)}{\sin(2x-1)}\right]$$
$$=\sin(2x-1)+2(x+1)\cos(2x-1)$$

Variables.—In accordance with established convention, the preceding examples have used x and y as the variable quantities. In engineering problems it very often happens that one of the variables is time measured from some selected instant. For example, the distance of a moving point from some fixed point may be denoted by s. The instant under consideration may be defined as t units of time later than some particular instant. Then the rate of change of s with t is the rate of change of the distance of the moving point from the fixed point, and this is what is known as the " velocity " of the moving point relative to the fixed point.

Hence, the differential coefficient of distance with respect to time—or $\frac{ds}{dt}$—is the velocity of the point. Similarly, the differential coefficient of velocity with respect to time—or $\frac{dv}{dt}$ where v represents velocity—is the acceleration of the point. Incidentally, since $v=\frac{ds}{dt}$, the acceleration $\frac{dv}{dt}$ is $\frac{d}{dt}\left(\frac{ds}{dt}\right)$, which is written $\frac{d^2s}{dt^2}$ and is the *second* differential coefficient of s with respect to t.

If the relation between s and t can be expressed mathematically, velocity and acceleration at any instant can thus be determined by means of the differential coefficients.

Example in Kinematics.—If in a reciprocating engine the crank-throw is r, and the length of the connecting rod is L, the distance of the cross-head pin from the centre-line of the crankshaft is (very nearly)

$$s=L-\frac{r^2}{4L}+r\cos(6{\cdot}28nt)+\frac{r^2}{4L}\cos(12{\cdot}56nt)\ldots \quad (9)$$

where r is the speed of the crankshaft in revolutions per second and t is the time in seconds from an instant at which the crank is on inner dead centre.

On differentiating this expression with respect to t, the quantity $L=\frac{r^2}{2L}$ disappears because, L and r being constant, their rate of change is zero. The other two terms are dealt with by use of (3), and

$$\frac{ds}{dt}=-6{\cdot}28nr\sin(6{\cdot}28nt)-12{\cdot}56\frac{nr^2}{4L}\sin(12{\cdot}56nt)\ldots \quad (10)$$

This represents the velocity of the cross-head pin at any instant defined by the value of t.

Differentiating (9) with respect to t,

$$\frac{d^2s}{dt^2}=-(6{\cdot}28n)^2r\cos(6{\cdot}38nt)-(12{\cdot}56n)^2\frac{r^2}{4L}\cos(12{\cdot}56nt) \quad (11)$$

This represents the acceleration of the cross-head pin at any instant defined by the value of t.

The simplest alternative method of determining velocity and acceleration of the cross-head pin is a graphical one (Fig. 154). The positions of the crank-pin, after suitable equal intervals of time, are marked on the circle representing its path. The corresponding positions of the cross-head pin are determined from these as centres by striking arcs (of radius representing the length of the connecting rod) intersecting the straight path of the cross-head pin. Consequently, the distance through which the cross-head pin moves during any one of the selected short intervals of time can be determined by direct measurement and thus the mean velocity over that interval of time is easily calculated. This may be taken as the velocity of the cross-head pin at the middle instant of the interval concerned.

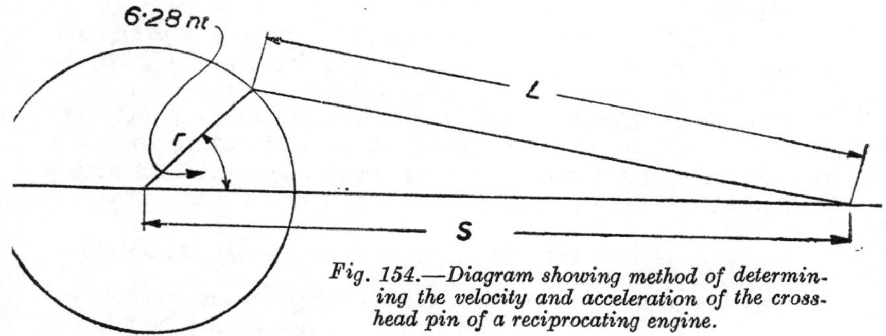

Fig. 154.—Diagram showing method of determining the velocity and acceleration of the cross-head pin of a reciprocating engine.

The acceleration at the beginning of any interval is the difference between the distance covered in that interval and that covered in the preceding interval, divided by the square of the interval.

The graphical method is thus simple in principle, but it has to be carried out with great accuracy if the result (particularly the figure for acceleration) is to be reliable.

Use of the calculus makes it easier to attain a high standard of accuracy.

Maxima and Minima.—When a quantity varies, it often does so in such a way as to attain one or more maximum or minimum values. For example, the velocity of a cross-head is zero at each end of its stroke and attains a maximum value somewhere near the middle of the stroke. If the changing values of any such quantity are expressed graphically (as, for example, variations of atmospheric pressure are traced by a recording barometer), the maximum and minimum values, and the positions they occupy, can be seen at a glance. If, however, the variations of the quantity can be expressed mathematically, the differential calculus may be used to ascertain maximum and minimum values and where they occur.

When a quantity is increasing, its value at any instant is greater than it was at a slightly earlier instant, so that the change in value between the earlier instant and the later one has been positive. The change in time from one instant to a later one is also positive. Consequently, the rate of change of the quantity with time is positive because it is equal to a positive quantity divided by a positive quantity.

Similarly, when one quantity is decreasing, whilst another one is increasing, the rate of change of the former with respect to the latter is negative.

If a quantity is increasing during one interval of time and decreasing during an immediately following interval, there will be some instant at which it is neither increasing nor decreasing, because it has just finished increasing and is just about to begin decreasing. At that instant the quantity has a maximum value.

Similarly an instant that lies at the end of a period of decrease and the beginning of a period of increase, marks a minimum value of the varying quantity.

At any such instant the rate of change (the differential coefficient) is neither positive nor negative, but is zero. It may be concluded, therefore, that when the differential coefficient of A with respect to B is zero, A has either a maximum or a minimum value. Conversely, to ascertain *when* A has a maximum or minimum value it is only necessary to determine when its differential coefficient is zero. To decide whether any one of these "stationary values" is a maximum or a minimum, a little more work has to be done.

In the neighbourhood of a maximum value the rate of a change is first positive, then zero at the maximum, and then negative. In other words, the rate of change is decreasing in the neighbourhood of a maximum, or again, the rate of change *of* the rate of change is negative at a maximum value. Conversely, the rate of change *of* the rate of change is positive at a minimum value.

In this way it is possible to decide whether any particular

value that makes the rate of change zero is a maximum or a minimum.

Example of Determination of Maxima and Minima.—These points may be illustrated by considering the previous "Example in kinematics." While the crankshaft rotates, the value of s, the distance from crankshaft to cross-head pin, is continually varying. Has it any maximum or minimum values ? The geometry of the mechanism makes it clear that the answer is "Yes," but it is instructive to consider how the differential calculus answers the question.

The variation of s with time will have a maximum or minimum value if $\frac{ds}{dt}$ ever becomes zero. Now the value of $\frac{ds}{dt}$ at any instant is given by equation (10) and for the easier understanding of what follows, the quantity sin $(12{\cdot}56nt)$ is replaced, according to a standard trigonometrical transformation, by 2 sin $(6{\cdot}28nt)$ cos $(6{\cdot}28nt)$. After further simplification (10) becomes

$$\frac{ds}{dt}=6{\cdot}28nr \sin (6{\cdot}28nt) \left[(1+\frac{r}{L} \cos (6{\cdot}28nt)\right] \quad . \; . \; . \; . \quad (12)$$

Similarly (11) becomes

$$\frac{d^2s}{dt^2}=-(6{\cdot}28n)^2r \left[\cos (6{\cdot}28nt)+\frac{r}{L} \cos (12{\cdot}56nt)\right] \quad . \; . \; . \quad (13)$$

Examination of (12) shows that $\frac{ds}{dt}=0$, when $\sin (6{\cdot}28nt)=0$, that is, when $6{\cdot}28nt=0$, $3{\cdot}14$, $2\times3{\cdot}14$, etc. or when

$$t=0, \frac{1}{2n}, \frac{1}{n}, \frac{3}{2n}, \text{etc.,}$$

and since n is the number of revolutions per second, $\frac{1}{n}$ is the number of seconds per revolution. Consequently the values of t correspond to 0, $\frac{1}{2}$, 1, $\frac{3}{2}$, etc., revolutions of the crank from the inner dead centre. In other words, the maximum and minimum distances of the cross-head from the crankshaft occur when the crank is at the inner and outer dead-centres.

To determine the values of these maxima and minima and to use the calculus method of showing which is which, the various values of $6{\cdot}28nt$ are inserted in (9) and (13), with the following results

$6{\cdot}28nt$	0	$3{\cdot}14$	$2\times3{\cdot}14$	$3\times3{\cdot}14$
$\sin(6{\cdot}28nt)$	0	0	0	0
$\cos(6{\cdot}28nt)$	1	-1	1	-1
$\cos(12{\cdot}56nt)$	1	1	1	1
s	$L+r$	$L-r$	$L+r$	$L-r$
$\frac{d^2s}{dt^2}$	$-(6{\cdot}28n)^2r \times\left(1+\frac{r}{L}\right)$	$-(6{\cdot}28n)^2r \times\left(-1+\frac{r}{L}\right)$	$-(6{\cdot}28n)^2r \times\left(1+\frac{r}{L}\right)$	$-(6{\cdot}28n)^2r \times\left(-1+\frac{r}{L}\right)$
Sign of $\frac{d^2s}{dt^2}$	$-$	$+$	$-$	$+$

As the nature of the mechanism is such that $\frac{r}{L}$ is less than 1, the sign of $(-1+\frac{r}{L})$ is negative.

It will be seen that the stationary values of s are alternately $L+r$ and $L-r$, and the signs of $\frac{d^2s}{dt^2}$ agree that these are maxima and minima. From the geometry of the mechanism it is obvious that these values are correct, but the calculus method of determining maxima and minima can often lead to results that could not be derived in any other way.

For example, suppose that it were required to know the maximum velocity of the crank-pin. This occurs (since velocity$=\frac{ds}{dt}$) when $\frac{d^2s}{dt^2}=0$ and from (13), this is the case when

$$\cos(6{\cdot}28nt)+\frac{r}{L}\cos(12{\cdot}56nt)=0 \quad . \quad . \quad . \quad . \quad (14)$$

By use of the trigonometrical identity $\cos 2A=2\cos^2A-1$, equation (14) is converted, after some rearrangement into

$$\cos^2(6{\cdot}28nt)+\frac{L}{2r}\cos(6{\cdot}28nt)-\tfrac{1}{2}=0$$

whence $\cos(6{\cdot}28nt)=-\frac{L}{4r}\pm\sqrt{\left(\left(\frac{L}{4r}\right)^2+\frac{1}{2}\right)}$ . . . (15)

Of the two values suggested by (15), the one involving two minus signs is inadmissible since it would make the cosine less than -1. The remaining value gives two values of $(6{\cdot}28nt)$ less than 360°, or in other words, two different angular positions of the crank corresponding to maximum or minimum velocities of the cross-head.

To take a numerical example, let $L=4r$,
when $\cos(6{\cdot}28nt)=-1+\sqrt{1{\cdot}5}=0{\cdot}2247$,
and from trigonometrical tables

$$6{\cdot}28nt=77°\ 1' \text{ or } -77°\ 1'.$$

To decide whether these angles give maximum or minimum velocities, the sign of the differential of $\frac{d^2s}{dt^2}$, i.e. $\frac{d^3s}{dt^3}$ is considered.

Differentiating (13) with respect to t,

$$\frac{d^3s}{dt^3}=(6{\cdot}28n)^3r\left[\sin(6{\cdot}28nt)+\frac{2r}{L}\sin(12{\cdot}56nt)\right]$$

When $6{\cdot}28nt=77°\ 1'$, both terms inside the square bracket are positive and therefore $\frac{d^3s}{dt^3}$ is positive, so that the velocity of the cross-head pin for that crank position is a minimum. As will be seen presently, the velocity is then negative (*i.e.* towards the crankshaft), so that an algebraic minimum velocity from the crankshaft means a numerical *maximum* velocity *towards* the crankshaft.

When $6{\cdot}28nt=-77°\ 1'$, both terms inside the square bracket are positive and therefore $\frac{d^3s}{dt^3}$ is positive, so that the velocity of the cross-head pin for that crank position is a maximum from the crankshaft.

The actual velocities are obtained by use of (12), thus—

$$\frac{ds}{dt}=-6{\cdot}28nr\sin 77°\ 1'\left[1+\frac{1}{4}\cos 77°\ 1'\right]$$
$$=-6{\cdot}36nr$$

and

$$\frac{ds}{dt}=-6{\cdot}28nr\sin(-77°\ 1')\left[1+\frac{1}{4}\cos(-77°\ 1')\right]$$
$$=6{\cdot}36nr$$

The two velocities are thus equal and opposite and are slightly greater than the crank-pin velocity which is $6{\cdot}28nr$. The maximum (maximum outward) and minimum (minimum inward) velocities occur when the cross-head is at a distance from the crankshaft centre line determined by use of (9), thus—

$$s=L-\frac{r^2}{4L}+r\cos 77^\circ\ 1'+\frac{r^2}{4L}\cos 154^\circ\ 2'$$

and writing $L=4r$, this becomes

$$s=4r-\frac{r}{16}+0{\cdot}414r+\frac{r}{16}(-0{\cdot}8094)$$
$$=4{\cdot}112r$$

The extreme distances of the cross-head pin from the crankshaft centre-line are $L-r=3r$, and $L+r=5r$, so that these maximum velocities occur not at mid-stroke (which is $s=4r$) but at a point farther away from the crankshaft than mid-stroke.

Calculus of Small Differences.—A knowledge of the differential coefficient of one quantity B with respect to another A makes it possible to calculate by how much B will change when A is changed by a small amount, or *vice versa*. The differential calculus is not essential for such calculations, but it is very often more accurate than the more elementary method.

For example, let a quantity y be connected with a quantity x by the relation—

$$y=6\sin x.$$

Now suppose that it were asked, " If y changes from the value 4 to the value 4·01, by how much does x change ? " To answer this problem directly it would be necessary first to find the original value of x from

$$\sin x=\frac{4}{6}$$

then to determine the second value of x from

$$\sin x=\frac{4{\cdot}01}{6}$$

and then to obtain the difference between the two values of x by subtraction. Trigonometrical tables would be needed and interpolation would be required. By use of the principles of the calculus, however, a formula can be derived for easy and more accurate solution of this problem without using trigono metrical tables.

Thus
$$\frac{dy}{dx}=6\cos x=6\sqrt{(1-\sin^2 x)}$$
$$=6\sqrt{\left(1-\left(\frac{y}{6}\right)^2\right)}=\sqrt{(36-y^2)}$$

Now $\frac{dy}{dx}$ is (very nearly) the ratio between any small change $\triangle y$ in y and the corresponding small change $\triangle x$ in x, *i.e.*—

$$\triangle y=\left(\frac{dy}{dx}\right)\triangle x$$

or

$$\triangle x=\frac{\triangle y}{\left(\frac{dy}{dx}\right)}=\frac{\triangle y}{\sqrt{(36-y^2)}}$$

If the initial value of y is 4 and the new value 4·01, then $y=4{\cdot}01-4=0{\cdot}01$, and so

$$\triangle x=\frac{0{\cdot}01}{\sqrt{(36-16)}}=0{\cdot}00224$$

In this formula the value given to the y in the denominator is the initial value 4. It is more accurate to use the value 4·005 half-way between the initial value 4 and the new value 4·01, but the gain in accuracy in this case is not worth the extra trouble involved in squaring 4·005 as compared with squaring 4.

This calculation is an example of the use of the "calculus of small differences," a branch of mathematics which makes use of the fact that although a differential coefficient is strictly the ratio between two *infinitely small* quantities, it is very nearly the same thing as the corresponding ratio between two quantities that are small but not infinitely small. Its advantage is that it shows how to calculate a difference directly instead of by the obvious, but less accurate, method.

CHAPTER XXVII

INTEGRATION

THE normal meaning of the term "to integrate" is "to make into one whole" and in mathematics the meaning is similar, but is perhaps more accurately defined as "to determine the whole from a knowledge of the relations between its infinitely small parts." This definition hints at the converse relation between integration and differentiation, which is a determination of the relations between infinitely small parts. In what

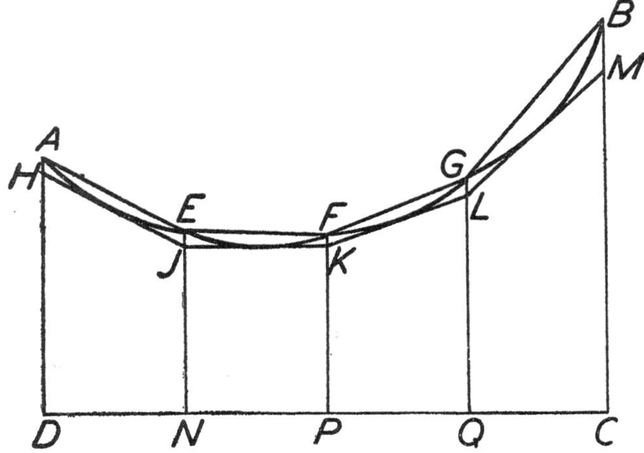

Fig. 155.—Area of irregular figure.

circumstances is it necessary to deduce the whole from consideration of its infinitely small parts? An important example may be understood by reference to Fig. 155.

Suppose that it is required to know the area of the figure $ABCD$, in which AD and BC are perpendicular to DC and AB is curved. An easy way of determining the approximate value of the area is to divide it into vertical strips by the lines EN, FP, and GQ, to join AE, EF, FG, and GB by straight lines and to add together the areas of the four strips. (The area of each strip is equal to its horizontal width multiplied by the average of the heights of its sides.) The area so determined is actually greater than the area required, because the lines AE, EF, FG, and GB all lie *above* the original curve AB.

Now suppose that a straight line HJ is drawn to touch the original curve at a point half-way between A and E, JK is drawn touching the curve between E and F and KL and LM are similarly located. Then the total area of the four strips bounded at their tips by HJ, JK, KL, and LM may be determined as before and this is *less* than the required area because HJ, JK, KL, and LM all lie *below* the curve AB.

It is clear from the diagram that the difference between the total area of the larger strips and the total area of the shorter strips is small and therefore that the area below the curve AB, although not *exactly* determined, is known within reasonably narrow limits of error. It is also fairly obvious that if the original figure were divided into (say) eight strips instead of four, the limits of error would be even narrower because the differences between curve, chord, and tangent are reduced when a shorter piece of the curve is considered. It can be said, in fact, that the difference between the total area of the strips and the area under the curve can be reduced to any desired amount by dividing the given figure into a sufficiently large number of strips. If an *infinitely* large number of strips be used, the difference becomes zero, or in other words, the method of adding areas of strips becomes accurate.

If there is a known mathematical relation between the ordinate of the curve (*i.e.* its height above DC) at any point and the horizontal distance of the point from AD, the integral calculus provides a means of adding together the areas of an infinite number of infinitely narrow strips—an operation that is otherwise impracticable. It is necessary to state here, however, that whilst the strict straightforward process of integration can often be carried out with no difficulty, and sometimes with a certain amount of difficulty, there are circumstances in which it cannot be carried out at all. In other words, there are mathematical functions that cannot be integrated accurately, and in such cases it may be essential to adopt the graphical method of determining the area, using any conveniently large number of strips for the purpose. There is thus a sharp distinction between differentiation and integration in that whilst every mathematical function can be differentiated, by following certain fairly simple rules, it is quite easy to write down any number of functions that, in the strict mathematical sense, cannot be integrated at all.

Relation between Integration and Differentiation.—In Fig. 156 the upper curve is one that is defined by a known mathematical relation (for example, $y=3x^2$) between the height y of the curve at a point distant x from AO. The point B is any typical one; $BD=y$ and OD is x. The area of the figure bounded by the straight lines AO, OD, and DB, and the curve itself between A and B is denoted by c. Since the area depends on the distance between AO and BD, the value of c depends on OD, or in other words, C is a function of x.

Now consider an ordinate CE, a little to the right of BD.

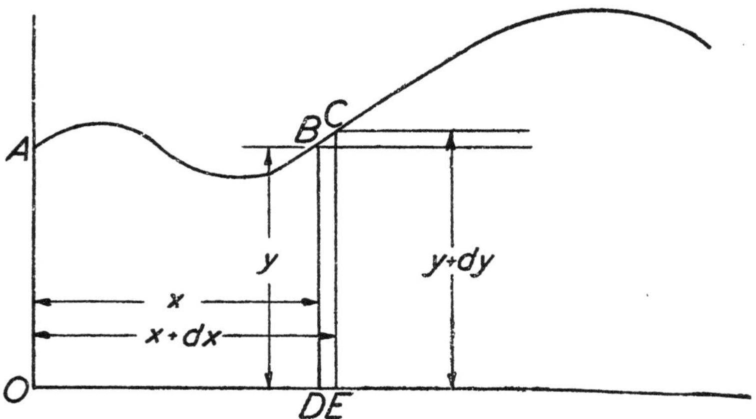

Fig. 156.—Curve of known relation.

The short distance DE may be regarded as a small change in x and denoted by dx. The height CE will be slightly different from BD and may be denoted by $y+dy$. The area under the curve between A and C is slightly greater than c and the difference may be denoted by dc. This difference is the area of the strip $DBCE$, whose width is dx and whose mean height is between y and $(y+dy)$. When BD and CE are infinitely close together, dy is infinitely small, so that y and $(y+dy)$ are equal, the mean height of the strip is y, and its area is ydx. Hence—

$$dc=ydx, \text{ or } \frac{dc}{dx}=y$$

The differential coefficient of c with respect to x is therefore y, but as the function c is at present unknown, whereas y is a known function of x, the relation $\frac{dc}{dx}=y$ expresses the fact that c is the function of x whose differential coefficient with respect to x is equal to y. The mathematical process of " integrating " the areas of the strips thus resolves itself into finding the function which, when differentiated, will give y. The relation between integration and differentiation is similar to that between division and multiplication, because when we divide P by Q, we find the quantity which when multiplied by Q will give P.

Symbol for Integration.—The area c under the curve AB is the sum of the areas of an infinitely large number of strips (whose individual areas are each $dc=ydx$) lying between the ordinate AO (for which $x=0$) and the ordinate BD (for which $x=OD$). This is expressed in mathematical shorthand by writing

$$c=\int_{x=0}^{x=OD} ydx$$

The symbol $\int$ means " the sum of." The symbol dx shows that x is the independent variable. This indication must always be given because the sign $\int$ is meaningless without it. The quantity y (which must in any particular case be represented as a function of x) is the quantity to be integrated. The notes $x=0$ and $x=OD$ indicate the " limits " between which the integration is to be carried out. The mathematical operations involved in determining c by this integral formula are—

(1) Determine the function of x whose differential coefficient with respect to x is y.
(2) Find the value of that function when $x=OD$.
(3) Find the value of that function when $x=0$.
(4) Subtract the second value from the first value.

The reason why this procedure is correct is that any ordinate of the curve is equal to the differential coefficient (with respect to x) of the area under the curve up to that ordinate.

Limits of Integration.—The last statement above is true irrespective of the position of the left-hand limit of the curve and consequently integration, interpreted as reversed differentiation, cannot determine the area under the curve unless both left- and right-hand limits are defined. This is in accordance with the fact that the addition of a constant to a function of x does not alter the differential coefficient of that function of x with respect to x, because the differential coefficient of a constant is zero. If the limits are not defined, the value of the integral must be expressed with the addition of an " arbitrary constant " whose value must be determined with the aid of some additional information.

In the following list of some standard integrals, the arbitrary contant is denoted in each case by C:

Some Standard Integrals—

$$\int (ax+b)^n dx = \frac{(ax+b)^{n+1}}{a(n+1)}+C \tag{16}$$

$$\int a \sin (nx+b)dx = \frac{-a \cos (nx+b)}{n}+C \tag{17}$$

$$\int a \cos (nx+b)dx = \frac{a \sin (nx+b)}{n}+C \tag{18}$$

$$\int a \sec^2(nx+b)dx = \frac{a \tan (nx+b)}{n}+C \tag{19}$$

$$\int a \tan (nx+b)dx = \frac{a}{n} \log_e \cos (nx+b)+C \tag{20}$$

$$\int ab^{(cx+d)}\, dx = \frac{ab^{(cx+d)}}{c \log_e b}+C \tag{21}$$

If u, v, w, etc., are functions of x—

$$\int (u+v+w+\text{etc.})dx = \int udx + \int vdx + \int wdx + \text{etc.} + C \tag{22}$$

$$\int uvdx = v\int udx - \int \left(\int udx\right)\frac{dv}{dx}dx + C) \tag{23}$$

For example, using (22), (17), and (18),

$$\int (3 \sin 2x + 2 \cos 2x)dx$$

$$= \int 3 \sin 2x\, dx + \int 2 \cos 2x\, dx$$

$$= \frac{-3 \cos 2x}{2} + C_1 + \frac{2 \sin 2x}{2} + C_2$$

$$= -\frac{3}{2} \cos 2x + \sin 2x + C$$

Each integral gives rise to an arbitrary constant (denoted by C_1 and C_2) but the sum of any number of arbitrary constants is only one arbitrary constant and this is denoted by C.

Using (16),

$$\int_{x=1}^{x=2} (-2x+1)^2 dx = \left[\frac{(-2x+1)^3}{(-2) x 3}\right]_{x=1}^{x=2}$$

$$= \frac{(-2 \times 2+1)^3}{-6} - \frac{(-2 \times 1+1)^3}{-6}$$

$$= 4 \cdot 5 - \tfrac{1}{6} = 4 \cdot 333$$

Equation (23) is occasionally useful in evaluating the integral of the product of two functions of x. It must be emphasised, however, that it is not certain to help in any particular case because it merely substitutes the integral of the product of two other functions of x for the original integral. It is useful only if the second term on the right-hand side is easier to evaluate than is $\int uvdx$.

It may be mentioned again that there are many mathematical functions that cannot be strictly integrated inasmuch as no function can be found on differentiation to give the original function. In such a case, however, the numerical value of the integral between specified limits may be found by plotting the curve of the function between those limits and determining the area under the curve by dividing it into strips, or by use of the planimeter. Here the approximate method gives an answer, whereas the exact method fails to work at all. What we may call the " separate strip " method always gives a result of adequate accuracy; sometimes the integral calculus can be used and when it can it is usually quicker and more accurate.

Moment of Inertia.—The energy possessed by a body by virtue of velocity is called its " kinetic energy " and is equal to half the mass multiplied by the square of its velocity. If a body is rotating about a fixed axis, different parts of it have different velocities, and consequently it is more difficult to calculate its kinetic energy. To take the simplest case—that of a disc rotating about its axis—points on the curved surface have the greatest velocity of any, whilst points on the axis have no velocity at all.

The velocity of any point is equal to the angular velocity of the disc multiplied by the distance of the point from the axis. Consequently, all points at any particular radius may be lumped together in calculating kinetic energy because they have a common velocity. Further, all points lying in a cylindrical ring (as shown in Fig. 157) whose inner and outer radii are r and $(r+dr)$ lie at nearly the same radius r if dr is very small. The volume of such a ring is equal to the circumference multiplied by the thickness multiplied by the width, and the mass is equal to that product multiplied by the density of the material, or

Mass$=2\pi\ r\ dr\ w\ s$, where s is the density.

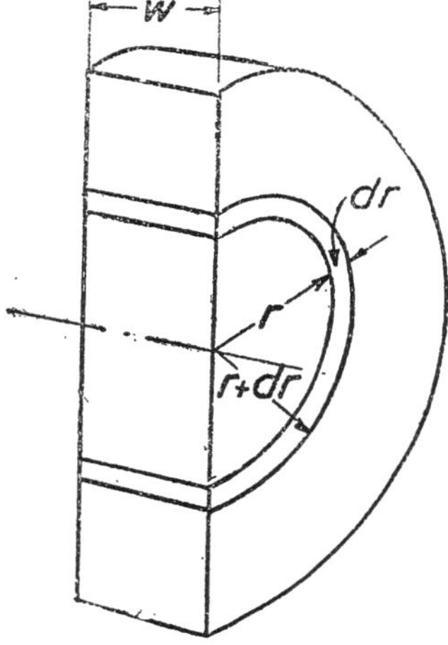

Fig. 157.—Cylindrical ring.

The velocity of the ring is equal to r times the angular velocity, assuming that the value r is taken as a sufficiently close approximation to the distances of the axis from the various particles of material which actually range from r to $(r+dr)$. If the angular velocity is A, the square of the velocity of the ring is A^2r^2, and the kinetic energy is

$$\tfrac{1}{2}\text{ Mass } x\ (\text{Velocity})^2=\pi\ A^2 s\ w\ r^3\ dr.$$

The same expression, with suitable values of w and r, applies to any ring. The values of π, A, and s are common to all rings. (In the particular case shown, that of a flat-sided disc, the value of w is also common to all rings.)

The total kinetic energy of the whole disc is the sum of the values of this expression for all rings, or

$$=A^2\ s\ w\ (\text{Sum of } r^3\ dr \text{ for all rings}).$$

The value of the quantity within the bracket may be estimated by dividing the disc into thin rings and adding together $r^3\,dr$ for all of them. For example, if the radius of the disc is 10, and it is decided to consider rings of thickness 1, then $dr=1$, and the calculation of r^3dr would proceed thus—

r	r^3	dr	r^3dr
0	0	1	0
1	1	1	1
2	8	1	8
3	27	1	27
4	64	1	64
5	125	1	125
6	216	1	216
7	343	1	343
8	512	1	512
9	729	1	729
			Total $r^3dr=2025$

On the other hand, if it were decided to use rings of thickness 2, then $dr=2$, and the calculation would proceed—

r	r^3	dr	r^3dr
0	0	2	0
2	8	2	16
4	64	2	128
6	216	2	432
8	512	2	1024
			Total $r^3dr=1600$

This gives a smaller result than before, because the steps are coarser. Thus for the ring whose inner and outer radii are 2 and 4, the value of $r^3\,dr$ is $2^3\times 2=16$. The corresponding value on the basis of $dr=1$ is derived from a ring for which $r=2$ and another for which $r=3$, and is

$$2^3\times 1+3^3\times 1=35.$$

It is true to say that the thinner the rings the greater will the total $r^3\,dr$ become, and it may be expected that the choice of infinitely small rings will yield a definite value for $r^3\,dr$ because that is actually the only perfectly correct procedure.

The integral calculus enables that summation to be effected because the expression $\int_{r=a}^{r=b} r^3\,dr$ means "the sum of the quantities $r^3\,dr$ between $r=a$ and $r=b$ when dr is infinitely small " and because there is a mathematical process for finding the value of that expression. In the special case just considered the maximum and minimum values (the limits) of r are 10 and 0, and so $b=10$ and $a=0$. Hence :

$$\int_{r=\ 0}^{r=10} r^3\,dr=\left[\frac{r^4}{4}\right]_{r=\ 0}^{r=10}=\frac{10000}{4}-\frac{0}{4}=2500$$

So the sum of $r^3\,dr$ calculated on $dr=2$ is 1600, on $dr=1$ it is 2025, and on $dr=\frac{1}{2}$ it will be something greater than 2025, whilst the integral calculus tells us (what we could not find by the simple arithmetical method) that if $dr=0$ the sum of $r^3\,dr=2500$, and this is the correct answer.

The quantity $\pi s\int w\,r^3\,dr$ is called the " moment of inertia " of the disc about its axis.

If w is constant for all rings, then $\int w\,r^3\,dr=w\int r^3\,dr$, and the integral is easily evaluated as was done above. If w is not constant but can be expressed in terms of r, it may, or may not, be possible to evaluate the integral according to the complexity of the expression wr^3. If it *is* possible, the value of the integral may be expressed as a general formula in terms of the outer radius of the disc or, in the case of a ring, in terms of its outer and inner radii. If it is *not* possible, the arithmetical method based on rings must be used, and the thinner the rings the more accurate will be the result.

Mean Values.—What is meant in general by a " mean value " is well illustrated by the case of mean velocity. If the velocity of a body varies over a given interval of time, the mean velocity for that period is equal to the total distance covered divided by the time. If the distance covered cannot be determined by actual measurement, it can be calculated, if the variation of velocity with time is known, by dividing the whole interval into very short intervals, multiplying the velocity at the middle of the short interval by the length of the short interval (this giving the distance covered during that interval), and adding together all the quantities so obtained. The result divided by the whole interval is the mean velocity during the interval.

Here again it may be surmised that the calculated value of the mean velocity will depend on the shortness of the intervals chosen for the basis of the calculation. The shorter (and therefore the more numerous) the intervals, the more accurate will be the result, and if the velocity can be expressed mathematically in terms of time the integral calculus will be able to give the mean velocity with perfect accuracy if the mathematical function is one that can be strictly integrated.

In general, if a quantity y depends on a quantity x, the mean value of y while x changes from a to b is

$$\left(\int_{x=a}^{x=b} y\,dx\right)\Big/(b-a)$$

For example, if $y=\frac{1}{2}x^3$, the mean value of y between $x=2$ and $x=4$ is

$$\left(\int_{x=2}^{x=4} \tfrac{1}{2}x^3\,dx\right)\Big/(4-2)$$

$$=\left(\left[\frac{x^4}{8}\right]_{x=2}^{x=4}\right)\Big/2=(32-2)/2=15$$

when $x=2$, $y=\frac{8}{2}=4$, and when $x=4$, $y=32$.

The arithmetic mean of the smallest and greatest values of y is $\frac{1}{2}(4+32)=18$, and this is seen to be greater than the true mean value of y.

As usual $\int_{x=a}^{x=b} y\,dx$ represents the area under the curve of y between the limits $x=a$ and $x=b$, and so even if the integral cannot be evaluated mathematically, it can be evaluated by drawing the curve and determining the appropriate area by any convenient method.

Mean Height of Sine Curve.—The sine curve is important in several branches of engineering, and it is instructive to consider the application of the integral calculus to the determination of its mean value. It is illustrated in Fig. 158, which shows a complete " wave-length " w of a sine curve whose " amplitude " is b. This means that the value of y varies between $+b$ and $-b$.

The equation of the curve shown, i.e. the relation between the x and y of any point on it, is

$$y=b\sin 2\pi\frac{x}{w}$$

(As x increases from 0 to w, y rises from 0 to 1, falls to -1, and rises again to 0.)

Now the mean height of the curve between any two points A and B whose distances from 0 are c and f is, according to the general formula,

$$\left(\int_{x=c}^{=} b \sin 2\pi\frac{x}{w}\,dx\right)\Big/(f-c)$$

The integration of $b \sin 2\pi\frac{x}{w}$ is a special case of (17) where $a=b$, $n=\frac{2\pi}{w}$, and $b=0$. As the limits $x=f$ and $x=c$ are defined,

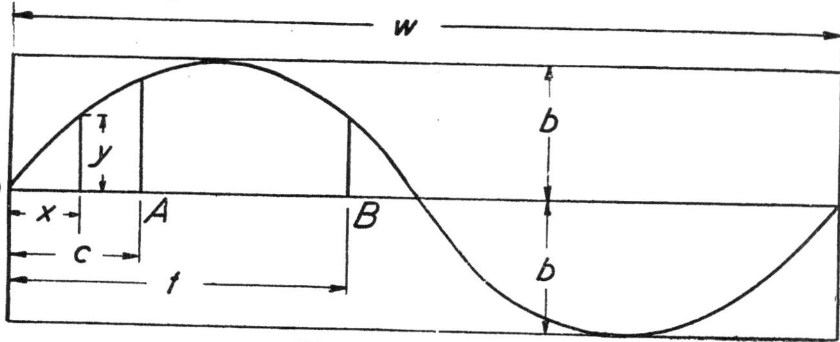

Fig. 158.—The sine curve.

no arbitrary constant appears. Therefore

$$\text{Mean height}=\left[-b\frac{w}{2\pi}\cos 2\pi\frac{x}{w}\right]_{x=c}^{x=f}\Big/(f-c)$$

$$=\frac{bw}{2\pi(f-c)}\left[\left(-\cos 2\pi\frac{f}{w}\right)-\left(-\cos 2\pi\frac{c}{w}\right)\right] \quad (24)$$

If the lower limit $c=0$, i.e. if A coincides with 0, the expression becomes simplified because the second term in the square bracket is then the cosine of zero, which is unity, and we have

$$H=\text{Mean height}=\frac{bw}{2\pi f}\left[1-\cos 2\pi\frac{f}{w}\right] \quad . \quad . \quad . \quad . \quad . \quad (25)$$

The following special values of f may be noted. If $f=w$, $H=0$. This is because in a complete wave-length there is as much positive as negative, and the total is nothing

If $\quad f=\frac{1}{2}w,\ H=\frac{2b}{\pi}=0{\cdot}637b$

If $\quad f=\frac{1}{4}w,\ H=\frac{2b}{\pi}=0{\cdot}637b$

These values of H are the same because the second quarter of the wavelength is the mirror image of the first quarter and the two quarters therefore have the same height.

It is interesting to apply the differential calculus to (25) in order to determine what length of the curve, starting from 0, will give the greatest mean height. In this operation f is the independent variable and H the dependent variable. Consequently it is necessary to differentiate H with respect to f, and this is an example of use of the formula (8), where

$$u=1-\cos 2\pi\frac{f}{w} \quad \text{and} \quad v=f$$

The quantity $\frac{bw}{\pi}$ is a multiplying constant and remains unchanged on differentiation.

In differentiating u, the differential coefficient of 1 is zero The second term is a special case of (3) where $a=-1$, $n=\frac{2\pi}{w}$ and $b=0$.

So
$$\frac{du}{df}=\frac{2\pi}{w}\sin 2\pi\frac{f}{w}$$

$$\frac{dv}{df}=1$$

Therefore :
$$\frac{dH}{df}=\frac{bw}{\pi}\left[f\frac{2\pi}{w}\sin 2\pi\frac{f}{w}-\left(1-\cos 2\pi\frac{f}{w}\right)\right]\Big/f^2$$

Using the trigonometrical transformations,

$$\sin\theta=2\sin\frac{\theta}{2}\cos\frac{\theta}{2}$$

and
$$1-\cos\theta=2\sin^2\frac{\theta}{2}$$

$$\frac{dH}{df}=2\frac{bw}{\pi}\sin\frac{\pi f}{w}\left[2\pi\frac{f}{w}\cos\pi\frac{f}{w}-\sin\frac{\pi f}{w}\right]$$

For a maximum value of H, the quantity in the square bracket is zero, and this means that

$$\tan\frac{\pi f}{w}=2\frac{\pi f}{w}$$

This equation cannot be solved directly, but examination of the trigonometrical tables shows that it is satisfied if

$$\frac{f}{w}=1{\cdot}162 \text{ approximately}$$

or if
$$f=0{\cdot}37w$$

Inserting this value of f in (25), it is found that the maximum value of the mean height is $H_{max}=\frac{b}{2\cdot324}[1+0\cdot684]=0\cdot723b$, occurring at a distance $0\cdot37w$ from the start.

If the values of H be plotted for a range of values of f, the curve is found to be a sine curve with constantly diminishing amplitude, and the value $H=0\cdot723b$ at $f=0\cdot37w$ is the crest of the first and highest wave.

MATHEMATICAL AND GENERAL CONSTANTS

1 foot-pound $=1\cdot3825\times10^7$ ergs.

1 horse-power $=33{,}000$ ft.-lb. per min.$=746$ watts.

1 atmosphere $=14\cdot7$ lb. per sq. in.$=2116$ lb. per sq. ft.$=760$ mm. of mercury$=10^6$ dynes per sq. cm. (nearly).

A column of water 2·3 ft. high is equivalent to a pressure of 1 lb. per sq. in.

The length of seconds pendulum$=39\cdot14$ in.

Joule's Equivalent to suit Regnault's H

is $\begin{cases} 744 \text{ ft.-lb.}=1 \text{ Fahr. unit.} \\ 1393 \text{ ft.-lb.}=1 \text{ Cent. unit.} \end{cases}$

Regnault's H$=606\cdot5+\cdot305\ t°$ C.$=1082+\cdot305\ t°$ F.

$pu^{1\cdot0646}=479.$

$\log_{10} p=6\cdot1007-\frac{B}{t}-\frac{C}{t^2}$

where $\log_{10} B=3\cdot1812$, $\log_{10} C=5\cdot0882$.

p is in lbs. per sq. in., t is absolute temperature Centigrade.

u is the volume in cu. ft. per lb. of steam.

Volts $\times$ amperes = watts.

1 electrical unit = 1000 watts-hours.

e = Base of Napierian (named after Napier, discoverer of logarithms) or natural hyperbolic Logarithms = 2·7182818.

To convert Napierian logarithms to common logarithms multiply by $\mu = (\cdot 4343)$.

To convert common to Napierian logs. multiply by $\frac{1}{\mu} = 2\cdot 30258509$, or 2·3026 approx.

$\pi = 3\cdot 14159265358979323846$

or 3·14159 approx. or 3·1416

or 3.142 or $3\frac{1}{7} = 3\cdot 142857$.

π correct to six decimal places $= \frac{355}{113}$.

Log $\pi = 0\cdot 49714987269413385435$.

The base of Napierian or hyperbolic logarithms is the sum of the series :

$$2 + \frac{1}{2} + \frac{1}{2\times 3} + \frac{1}{2\times 3\times 4} + \frac{1}{2\times 3\times 4\times 5} + \ldots$$

$g = 32\cdot 182\, f/s^2 = 980\cdot 9$ cms.$/s^2$. $= 32\cdot 2$ approx.

1 gallon = ·1606 cu. ft. = 10 lb. of water at 62° F.

1 pint of water = $1\frac{1}{4}$ lb.

1 cu. ft. of water weighs 62·3 lb. (1000 oz. approx.).

1 cu. ft. of air at 0° C. and 1 atmosphere weighs ·00559 lb.

1 radian $= \frac{180}{\pi} = 57\cdot 2958$ degrees.

π radians = 180°. $\frac{\pi}{2}$ radians = 90°.

Approximation

$(1\pm a)^2=1\pm 2a$

$(1\pm a)^3=1\pm 3a$

$\frac{1}{1\pm a}=1\mp a$

$\frac{1}{(1\pm a)^2}=1\mp 2a$

$\frac{1}{(1\pm a)^3}=1\mp 3a$

$\frac{1+a}{1-a}=1+2a$

$\frac{1-a}{1+a}=1-2a$

$\sqrt{1\pm a}=1\pm\frac{1}{2}a$

$\sqrt[3]{1\pm a}=1\pm\frac{1}{3}a$

a=small quantity

Trigonometry

$\sin^2 A+\cos^2 A=1$

$\sec^2 A=1+\tan^2 A$

$\operatorname{cosec}^2 A=1+\cot^2 A$

$\cos\frac{A}{2}=\sqrt{\frac{s(s-a)}{bc}}$

$\sin A/2=\sqrt{\frac{(s-b)\,(s-c)}{bc}}$

$\tan A/2=\sqrt{\frac{(s-b)\,(s-c)}{s(s-a)}}$

$\frac{\sin A}{a}=\frac{\sin B}{b}=\frac{\sin C}{c}=\frac{1}{2R}$

$\tan\frac{B-C}{2}=\frac{b-c}{b+c}\cot A/2$

$c^2=a^2+b^2-2ab\cos C$

$1+\tan^2 A=\sec^2 A$

Calculus

$\frac{d}{dx}(ae^x)=ae^x$	$\int ae^x dx=ae^x$
$\frac{d}{dx}(ax^n)=anx^{n-1}$	$\int ax^n dx=\frac{a}{n+1}x^{n+1}$
$\frac{d}{dx}(a\log_n x)=\frac{a}{x}\log_n e$	$\int\frac{a}{x}dx=a\log_e x$
$\frac{d}{dx}(a\sin bx)=ab\cos bx$	$\int a\sin bx\,dx=-\frac{a}{b}\cos bx$
$\frac{d}{dx}(a\cos bx)=-ab\sin bx$	$\int a\cos bx\,dx=\frac{a}{b}\sin bx$
$\frac{d}{dx}(a\tan bx)=ab\sec^2 bx$	$\int a\tan bx\,dx=\frac{a}{b}\log\sec bx$

The formulæ in the first column remain true when x is replaced by $x+C$, where C is any constant.

PULLEYS

Note.—$\frac{V^1}{V}$ = Ratio of distances moved by W and F.

Single Fixed Pulley

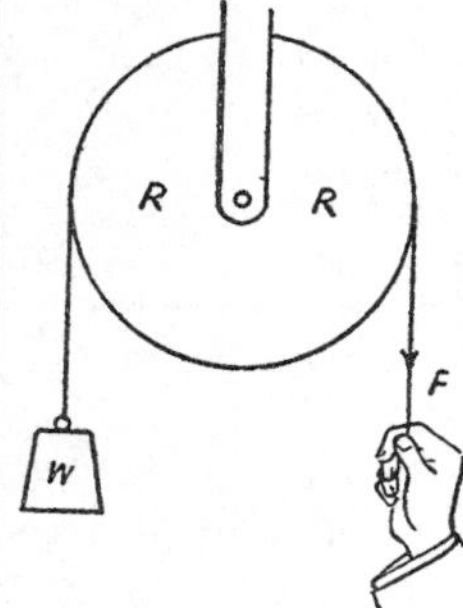

F:W = R:R
or F = W
$\frac{V^1}{V} = 1$

Single Movable Pulley

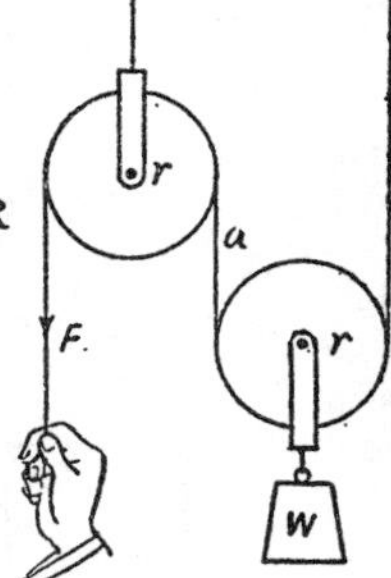

F:W = r:2r
or F = ½W
Note.—If the force is applied at "a" and acts upward, the result will be the same.
$\frac{V^1}{V} = 2$

Double Movable Pulley

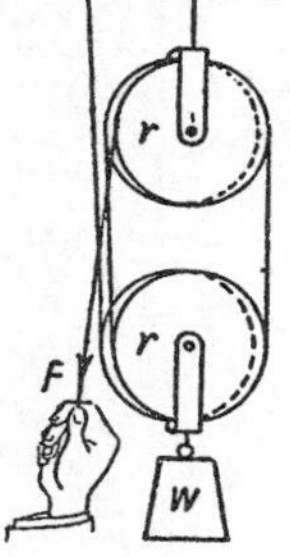

F:W=r:4r
or F = ¼W
$\frac{V^1}{V} = 4$

Double Movable Pulley

F = ¼W
$\frac{V^1}{V} = 4$

Multiple Movable Pulley

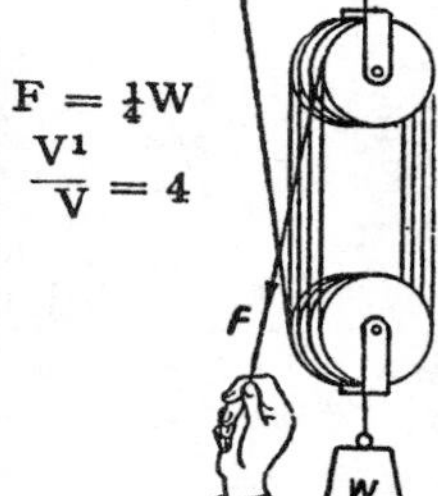

If "n" =any number of movable pulleys:
$F = \frac{W}{2n}$
$\frac{V^1}{V} = 2n$

Compound Pulleys

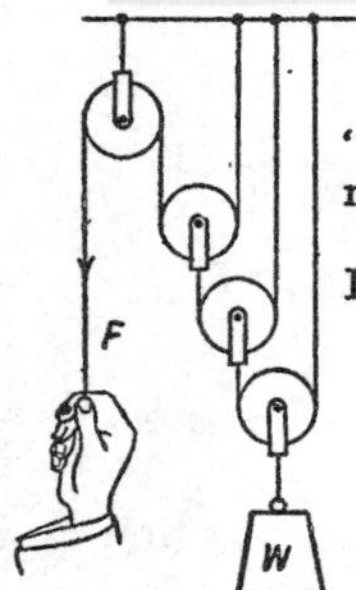

"n" = number of movable pulleys
$F = \frac{W}{2^n}$: $W = 2^nF$
$\frac{V^1}{V} = 2^n$

Oblique Fixed Pulley

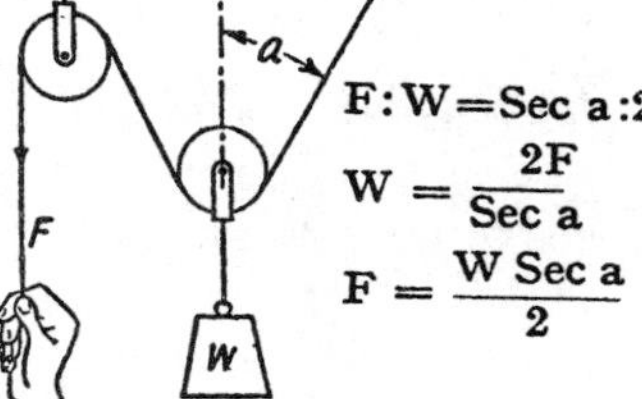

F:W=Sec a:2
$W = \frac{2F}{\text{Sec a}}$
$F = \frac{\text{W Sec a}}{2}$

Figs. 159 to 165

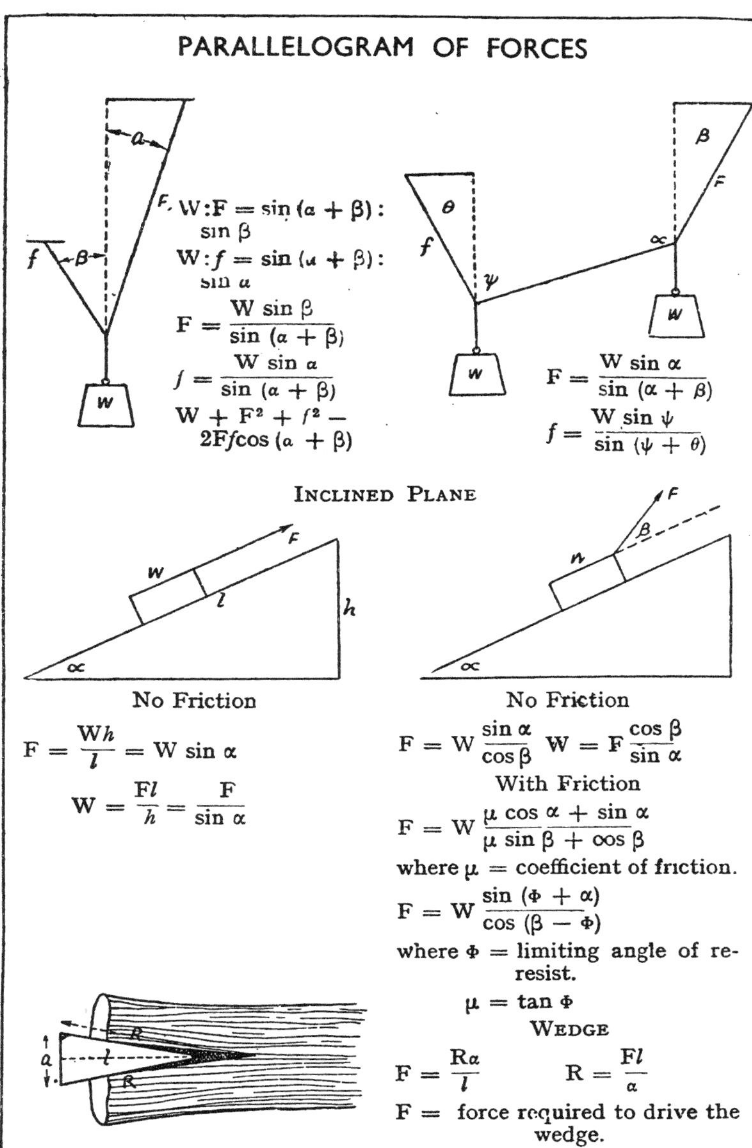

PARALLELOGRAM OF FORCES
$W : F = \sin(\alpha + \beta) : \sin\beta$
$W : f = \sin(\alpha + \beta) : \sin\alpha$
$F = \dfrac{W \sin\beta}{\sin(\alpha + \beta)}$
$f = \dfrac{W \sin\alpha}{\sin(\alpha + \beta)}$
$W + F^2 + f^2 - 2Ff\cos(\alpha + \beta)$
$F = \dfrac{W \sin\alpha}{\sin(\alpha + \beta)}$
$f = \dfrac{W \sin\psi}{\sin(\psi + \theta)}$
INCLINED PLANE
No Friction
$F = \dfrac{Wh}{l} = W \sin\alpha$
$W = \dfrac{Fl}{h} = \dfrac{F}{\sin\alpha}$
No Friction
$F = W\dfrac{\sin\alpha}{\cos\beta}$ $W = F\dfrac{\cos\beta}{\sin\alpha}$
With Friction
$F = W\dfrac{\mu\cos\alpha + \sin\alpha}{\mu\sin\beta + \cos\beta}$
where μ = coefficient of friction.
$F = W\dfrac{\sin(\Phi + \alpha)}{\cos(\beta - \Phi)}$
where Φ = limiting angle of re-resist.
$\mu = \tan\Phi$
WEDGE
$F = \dfrac{Ra}{l}$ $R = \dfrac{Fl}{a}$
F = force required to drive the wedge.

Figs. 166 to 170.

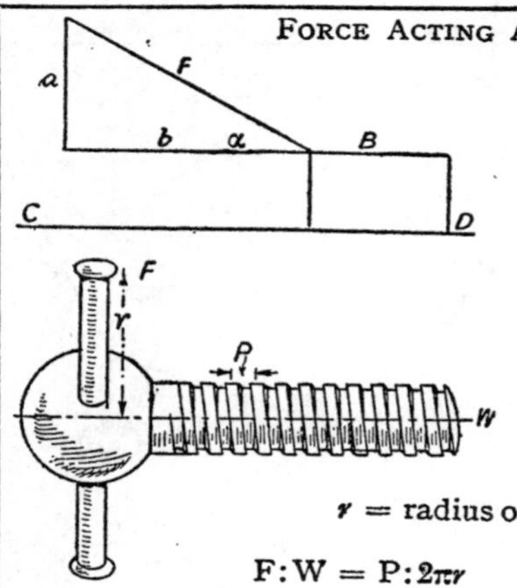

Force Acting at an Angle

Let line F represent the magnitude and direction of a force acting at an angle α to move the body B on line CD. Then the line a represents a part of F which presses the body B against CD. The line b represents the magnitude of the force which actually moves the body B.

$b = \sqrt{F^2 - a^2}$ $\quad$ $b = F \cos a$

Force by a Screw

P = pitch of screw (distance between threads)

r = radius on which force F acts. $\pi = 3{\cdot}1416$.

$$F:W = P:2\pi r \qquad F = \frac{WP}{2\pi r} \qquad W = \frac{F2\pi r}{P}$$

Figs. 171 and 172. Formulae relative to force acting at an angle and force by a screw

PENDULUM

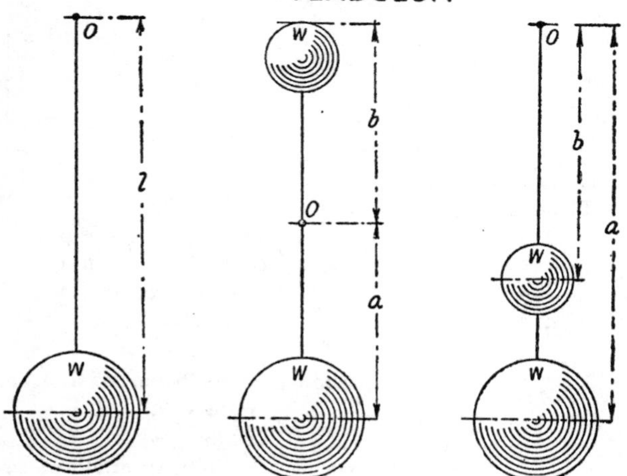

Figs. 173 to 175.

The Simple Pendulum, $t = \pi\sqrt{\frac{l}{g}}$, where l = length of pendulum in inches; t = time in secs. of one oscillation, $g = 32{\cdot}097$ ft. per sec. or $385{\cdot}163$ in per sec. $t = 2\pi\sqrt{\frac{l}{g}}$ for 25 swings.

The Compound Pendulum, in which o = centre of suspension and l = equivalent length of simple pendulum to give same time of oscillation, $l = \frac{a^2W + b^2w}{aW - bw}$

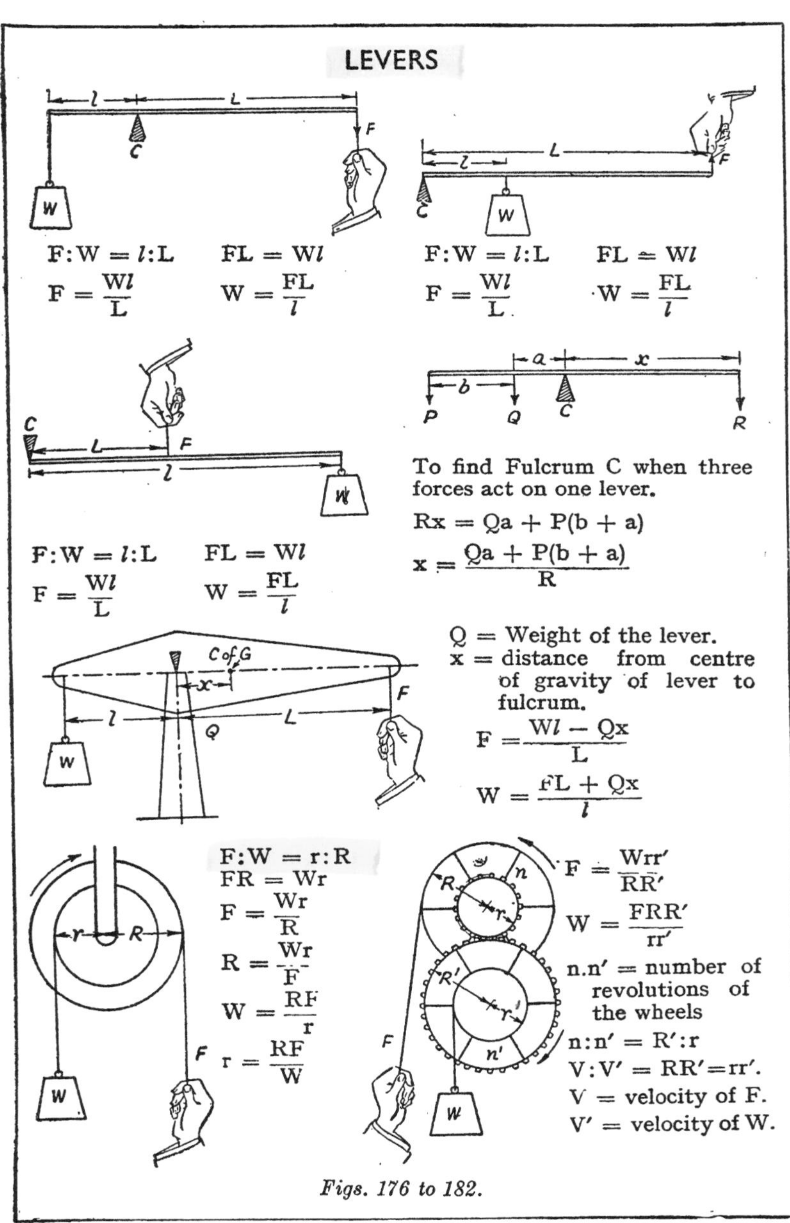

LEVERS
l
L
C
F
W
F:W = l:L
FL = Wl
F = Wl/L
W = FL/l
l
L
F
C
W
F:W = l:L
FL = Wl
F = Wl/L
W = FL/l
C
L
F
l
W
F:W = l:L
FL = Wl
F = Wl/L
W = FL/l
a
x
b
P
Q
C
R
To find Fulcrum C when three forces act on one lever.
Rx = Qa + P(b + a)
x = (Qa + P(b + a))/R
C of G
x
F
l
Q
L
W
Q = Weight of the lever.
x = distance from centre of gravity of lever to fulcrum.
F = (Wl − Qx)/L
W = (FL + Qx)/l
r
R
W
F
F:W = r:R
FR = Wr
F = Wr/R
R = Wr/F
W = RF/r
r = RF/W
R
n
r
R′
r′
n′
F
W
F = Wrr′/RR′
W = FRR′/rr′
n.n′ = number of revolutions of the wheels
n:n′ = R′:r
V:V′ = RR′=rr′.
V = velocity of F.
V′ = velocity of W.

Figs. 176 to 182.

CENTRIFUGAL FORCE

F = Centrifugal force in pounds
M = Mass or weight of revolving body in pounds.
v = Velocity of revol. body in ft. per sec.
R = Rad. of circle in which body revolves (ft.).

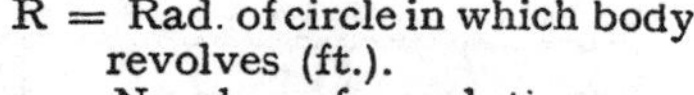

n = Number of revolutions per minute.
g = Coefficient of terrestrial Acceleration = 32·2.
π = 3·1416.

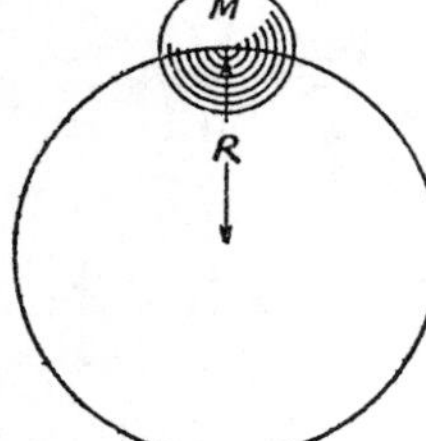

$$F = \frac{Mv^2}{gR} = \frac{Mv^2}{32{\cdot}2R} \qquad F = \frac{4MR\pi^2n^2}{60^2g} = \frac{MRn^2}{2933}$$

$$M = \frac{FgR}{v^2} = \frac{2933F}{Rn^2} \qquad R = \frac{Mv^2}{Fg} = \frac{2933F}{Mn^2}$$

$$n = \sqrt{\frac{2933F}{MR}} \qquad v = \sqrt{\frac{FRg}{M}}$$

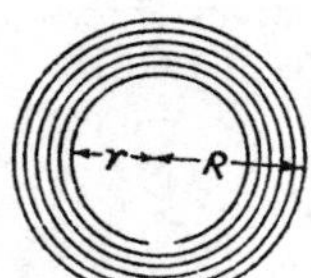

Centrifugal Tension (lbs.) of a Ring

$$\frac{Mn^2\sqrt{(R^2+r^2)}}{4150}$$

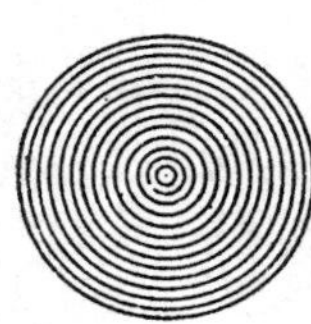

Centrifugal Tension of a Grindstone. Circle-plane, Cylinder rotating round its centre.

$$\frac{MRn^2}{4150}$$

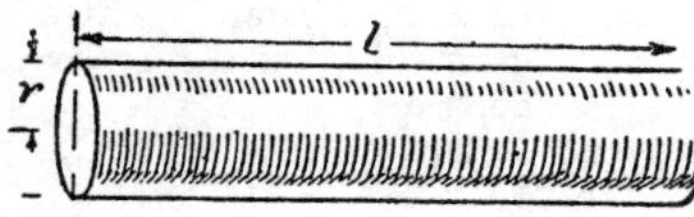

Centrifugal Tension of a Cylinder rotating round the diameter of its base.

$$\frac{Mn^2\sqrt{4l^2+3r^2}}{10260}$$

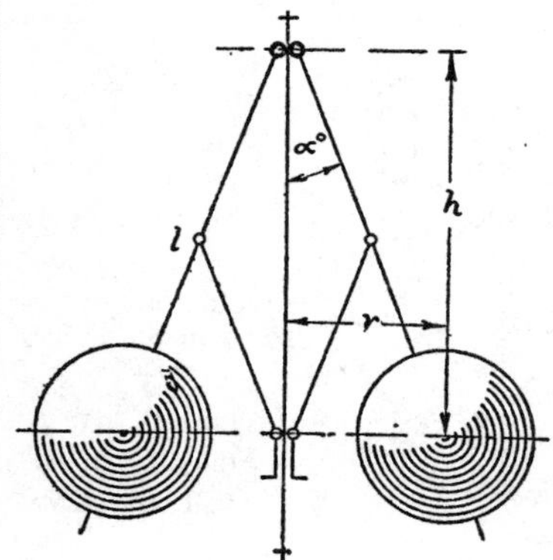

GOVERNOR

$$n = \frac{60}{2\pi}\sqrt{\frac{g}{h}} = \frac{54{\cdot}16}{\sqrt{h}} = \frac{54{\cdot}16}{\sqrt{l\cos a}}$$

$$h = \frac{2933}{n^2} \qquad l = \frac{2933}{n^2\cos a} = \frac{h}{\cos a}$$

$$\cos a = \frac{2933}{n^2l} = \frac{h}{l} \qquad r = \sqrt{l^2-h^2}$$

Figs. 183 to 187.

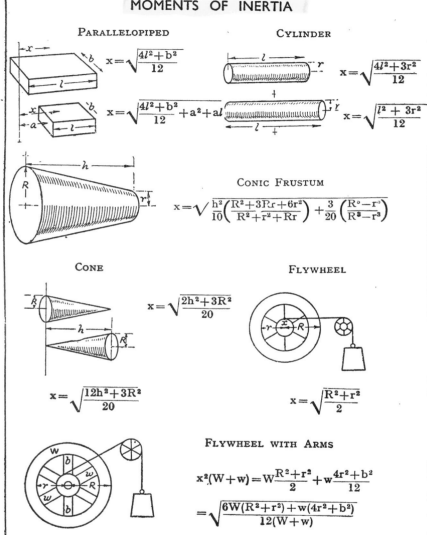

Parallelopiped

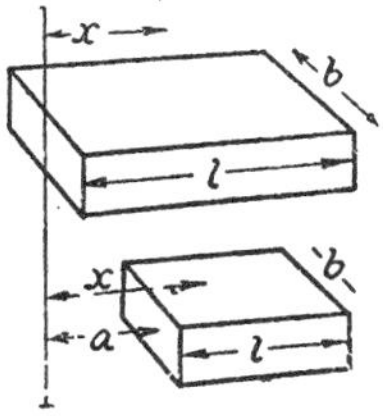

$$x=\sqrt{\frac{4l^2+b^2}{12}}$$

$$x=\sqrt{\frac{4l^2+b^2}{12}+a^2+al}$$

Cylinder

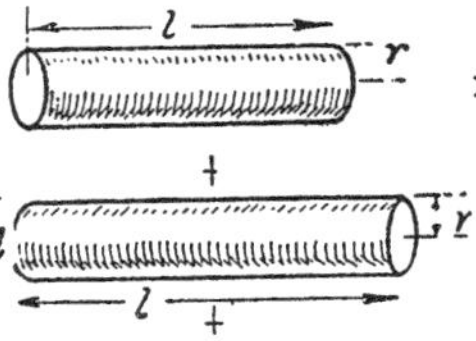

$$x=\sqrt{\frac{4l^2+3r^2}{12}}$$

$$x=\sqrt{\frac{l^2+3r^2}{12}}$$

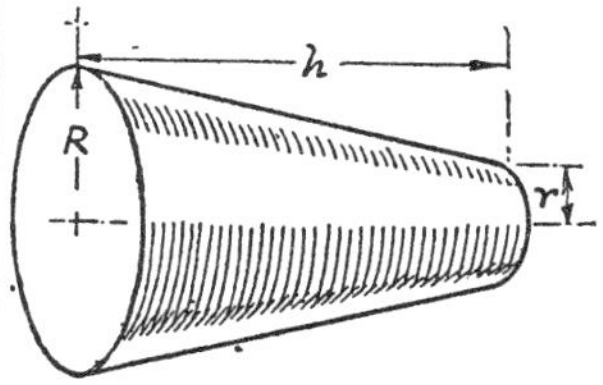

Conic Frustum

$$x=\sqrt{\frac{h^2}{10}\left(\frac{R^2+3Rr+6r^2}{R^2+r^2+Rr}\right)+\frac{3}{20}\left(\frac{R^5-r^5}{R^3-r^3}\right)}$$

Cone

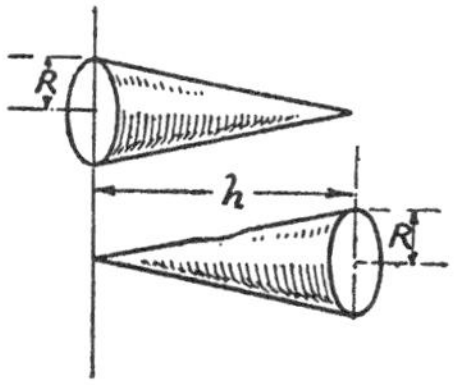

$$x=\sqrt{\frac{2h^2+3R^2}{20}}$$

$$x=\sqrt{\frac{12h^2+3R^2}{20}}$$

Flywheel

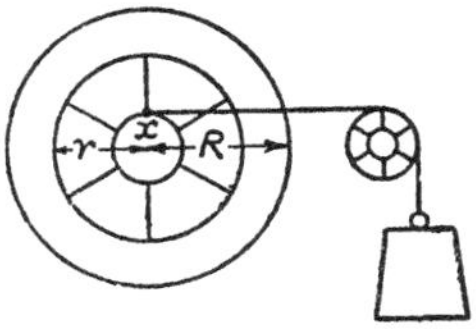

$$x=\sqrt{\frac{R^2+r^2}{2}}$$

Flywheel with Arms

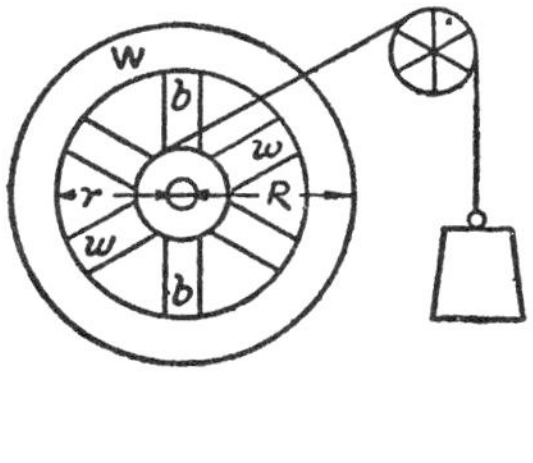

$$x^2(W+w)=W\frac{R^2+r^2}{2}+w\frac{4r^2+b^2}{12}$$

$$=\sqrt{\frac{6W(R^2+r^2)+w(4r^2+b^2)}{12(W+w)}}$$

Figs. 188 to 195.

THE GREEK ALPHABET

The Greek alphabet is as follows :

A α (alpha)	N ν (nu)
B β (beta)	Ξ ξ (xi)
Γ γ (gamma)	O o (omicron)
Δ δ (delta)	Π π (pi)
E ϵ (epsilon)	P ρ (rho)
Z ζ (zeta)	Σ σ (sigma)
H η (eta)	T τ (tau)
Θ θ (theta)	Υ υ (upsilon)
I ι (iota)	Φ ϕ (phi)
K κ (kappa)	X χ (chi)
Λ λ (lambda)	Ψ ψ (psi)
M μ (mu)	Ω ω (omega)

A USEFUL SERIES

$$=1+x+\frac{x^2}{2!}+\frac{x^3}{3!}+\frac{x^4}{4!}+\ \ldots\ldots$$

$$\log_e\frac{m}{n}=2\left[\left(\frac{m-n}{m+n}\right)+\frac{1}{3}\left(\frac{m-n}{m+n}\right)^3+\frac{1}{5}\left(\frac{m-n}{m+n}\right)^5+\ \ldots\ldots\right]$$

$$\log_e x=2\left[\left(\frac{x-1}{x+1}\right)+\frac{1}{3}\left(\frac{x-1}{x+1}\right)^3+\frac{1}{5}\left(\frac{x-1}{x+1}\right)^5+\ \ldots\ldots\right]$$

$$\sin\theta=\theta-\frac{\theta^3}{3!}+\frac{\theta^5}{5!}-\frac{\theta^7}{7!}+\ \ldots\ldots$$

$$\cos\theta=1-\frac{\theta^2}{2!}+\frac{\theta^4}{4!}-\frac{\theta^6}{6!}+\ \ldots\ldots$$

$$\theta=\tan\theta-\frac{\tan^3\theta}{3}+\frac{\tan^5\theta}{5}-\frac{\tan^7\theta}{7}+\ \ldots\ldots$$

POWERS AND ROOTS OF π AND g							
n	$\frac{1}{n}$	n^2	n^3	$\sqrt{n}$	$\frac{1}{\sqrt{n}}$	$\sqrt[3]{n}$	$\frac{1}{\sqrt[3]{n}}$
$\pi=$ 3·142	0·318	9·870	31·006	1·772	0·564	1·465	0·683
$2\pi=$ 6·283	0·159	39·478	248·050	2·507	0·399	1·845	0·542
$\frac{\pi}{2}=$ 1·571	0·637	2·467	3·878	1·253	0·798	1·162	0·860
$\frac{\pi}{3}=$ 1·047	0·955	1·097	1·148	1·023	0·977	1·016	0·985
$\frac{4}{3}\pi=$ 4·189	0·239	17·546	73·496	2·047	0·489	1·612	0·622
$\frac{\pi}{4}=$ 0·785	1·274	0·617	0·484	0·886	1·128	0·923	1·084
$\frac{\pi}{6}=$ 0·524	1·910	0·274	0·144	0·724	1·382	0·806	1·241
$\pi^2=$ 9·870	0·101	97·409	961·390	3·142	0·318	2·145	0·466
$\pi^3=$ 31·066	0·032	961·390	29,809·910	5·568	1·796	3·142	0·318
$\frac{\pi}{32}=$ 0·098	10·186	0·0095	0·001	0·313	3·192	0·461	2·168
$g=$32·2	0·031	1036·84	33,386·24	5·674	0·176	3·181	3·314
$2g=$64·4	0·015	4147·36	267,090	8·025	0·125	4·007	0·249

ENGLISH WEIGHTS AND MEASURES

Long Measure

·001 in.	=1 mil
$2\frac{1}{4}$ inches	=1 nail
4 inches	=1 hand
12 inches (in.)	=1 foot (ft.)
3 feet	=1 yard (yd.)
$5\frac{1}{2}$ yards	=1 rod, pole, or perch
40 poles (220 yards)	=1 furlong (furl.)
8 furlongs (1760 yards)	=1 mile (m.)
3 miles	=1 league
1 chain	=100 links (22 yards)
10 chains	=1 furlong
6 feet	=1 fathom
6080 feet or 1·1516 statute mile	=1 nautical mile

Area

(*Square Measure*)

144 square inches	=1 square foot
9 square feet	=1 square yard
$30\frac{1}{4}$ square yards	=1 square pole
40 square poles	=1 rood
4 roods	=1 acre (4840 sq. yards)
640 acres	=1 square mile

Measure of Capacity

(*Liquid or Dry Measure*)

4 gills	=1 pint
2 pints	=1 quart
2 quarts	=1 pottle
2 pottles	=1 gallon
4 quarts	=1 gallon
2 gallons	=1 peck
4 pecks	=1 bushel
3 bushels	=1 bag
4 bushels	=1 coombe
8 bushels	=1 quarter
12 bags	=1 chauldron
5 quarters	=1 load or wey
2 loads or weys	=1 last

Measures of Volume and Capacity

(*Cubic Measure*)

1728 cubic inches	= 1 cubic foot
27 cubic feet	= 1 cubic yard
1 marine ton	= 40 cubic feet
1 stack	=108 cubic feet
1 cord	=128 cubic feet

Wine Measure

4 gills	=1 pint
2 pints	=1 quart
4 quarts	=1 gallon
10 gallons	=1 anker
18 gallons	=1 runlet or rudlet
$31\frac{1}{2}$ gallons	=1 barrel
42 gallons	=1 tierce
63 gallons	=1 hogshead
2 tierces	=1 puncheon
$1\frac{1}{2}$ puncheons	=1 pipe or butt
2 pipes	=1 tun

Ale and Beer Measure

4 gills	=1 pint
2 pints	=1 quart
4 quarts	=1 gallon
9 gallons	=1 firkin
2 firkins	=1 kilderkin
2 kilderkins	=1 barrel
$1\frac{1}{2}$ barrels	=1 hogshead
$1\frac{1}{3}$ hogsheads	=1 puncheon
$1\frac{1}{2}$ puncheons or 2 hogsheads	=1 butt or pipe
2 pipes	=1 tun

Troy Weight

3·17 grains	=1 carat
24 grains	=1 pennyweight (dwt.)
20 pennyweights	=1 ounce (oz. troy)
12 ounces (troy)	=1 pound (troy)
1 pound (troy)	=5760 grains
1 pound (avoirdupois)	=7000 grains (troy)

To convert to troy avoirdupois × 1·097.

Avoirdupois Weight

27·34375 grains	=1 dram
16 drams	=1 ounce (oz.)
16 ounces	=1 pound (lb.)
14 pounds	=1 stone
2 stones (28 lbs.)	=1 quarter
100 lbs.	=1 cental
4 quarters	=1 hundredweight (cwt.)
20 cwt.	=1 ton

To convert avoirdupois to troy ÷ ·9115.

Apothecaries' Weight

20 grains or minims	=1 scruple
3 scruples	=1 drachm
8 drachms	=1 ounce
12 ounces	=1 pound

Apothecaries' Fluid Measure

60 minims	=1 fluid drachm
8 drachms	=1 fluid ounce
20 ounces	=1 pint (pt. or Octarius)
8 pints	=1 gallon (gal., C., or Congius)

Diamond and Pearl Weight

3·17 grains	=1 carat, or
4 pearl grains	=1 carat
$151\frac{1}{2}$ carats	=1 ounce (troy)

Paper Measure

24 sheets	=1 quire
20 quires	=1 ream
2 reams	=1 bundle
10 reams	=1 bale

UNITS AND EQUIVALENTS

One ft. lb.	=1 lb. raised 1 foot high
One B.T.U.	=1055 joules
One B.T.U.	=778·8 ft. lbs.
1 watt	$=10^7$ ergs per second
1 watt	=23·731 foot poundals per sec.
1 watt	=0·7376 ft. lb. per second
1 watt	=0·001341 h.p.
One H.P. hour	=0·746 kW. hour
One H.P. hour	=1,980,000 ft. lbs.
One H.P. hour	=2·545 B.T.U.s
One kw.H. (kilowatt hour)	=2,654,200 ft. lbs.
One kw.H.	=1000 watt hours
One kw.H.	=1·34 H.P. hours
One kw.H.	=3412 B.T.U.s
One kw.H.	=3,600,000 joules
One kw.H.	=859,975 calories
One H.P.	=746 watts
One H.P.	=0·746 kW.
One H.P.	=33,000 ft. lbs. per minute
One H.P.	=550 ft. lbs. per second
One H.P.	=2545 B.T.U.s per hour
One H.P.	=42·4 B.T.U.s per minute
One H.P.	=0·707 B.T.U.s per second
One H.P.	=178,122 calories per second

Mensuration

A and a=area.
b=base.
C and c=circumference.
D and d=diameter.
h=height.
$n°$=number of degrees.
p=perpendicular.
R and r=radius.
s=span or chord.
v=versed sine.

Square

$a=\text{side}^2$; $\text{side}=\sqrt{a}$; $\text{diagonal}=\text{side}\times\sqrt{2}$.
Side × 1·4142=diameter of circumscribing circle.
Side × 4·443=circumference of circumscribing circle.
Side × 1·128=diameter of circle of equal area.
Area in square inches × 1·273=area of equal circle.

Circle

$a=\pi r^2=d\frac{2\pi}{4}=\cdot 7854d^2=\cdot 5cr.$

$c=2\pi r$, or $\pi d=3\cdot 14159d$, or $3\cdot 1416d$ approx.$=3\cdot 54\sqrt{a}=3\frac{1}{7}d$ or $\frac{22}{7}d$ approx.

Side of equal square=$\cdot 8862d$.
Side of inscribed square=$\cdot 7071d$.
$d=\cdot 3183c$.
Circle has maximum area for given perimeter.

Annulus

$a=(D+d)\ (D-d)\ \frac{\pi}{4}=\frac{\pi}{4}(D^2-d^2).$

Segment of Circle

a=area of sector−area of triangle

$=\frac{4v}{3}\sqrt{(0\cdot 625v)^2+(\frac{1}{2}s)^2}.$

Length of Arc

Length of arc $= \cdot 0174533 n^\circ r$

$$= \frac{1}{3}\left(8\sqrt{\frac{s^2}{4}+v^2}-s\right)$$

Approx. length of arc $= \frac{1}{3}$(8 × chord of $\frac{1}{2}$ arc − chord of whole arc).

$$d = \frac{(\frac{1}{2}\text{ chord})}{v} + v$$

radius of curve $= \dfrac{s^2}{8v} + \dfrac{v}{2}$

Sector of Circle

$a = \cdot 5r \times$ length of arc $= \dfrac{n^\circ}{360} \times$ area of circle.

Rectangle or Parallelogram

$$a = bp.$$

Trapezium

(Two Sides Parallel)

$a =$ mean length of parallel sides × distance between them.

Ellipse

$a = \dfrac{\pi}{4} \mathrm{D}d = \pi \mathrm{R}r.$

$c = \sqrt{\dfrac{\mathrm{D}^2 + d^2}{2}} \times \pi$ approx., or $\pi \dfrac{\mathrm{D}a}{2}$ approx.

Parabola

$$a = \frac{2}{3} bh.$$

Cone or Pyramid

Surface $= \dfrac{\text{circum. of base} \times \text{slant length}}{2} +$ area of base.

Volume = area of base × $\frac{1}{3}$ vertical height

Frustum of Cone

Surface $=(\mathrm{C}+c)\times\frac{1}{2}$ slant height $+$ area of ends.
Volume $=\cdot 2618h(\mathrm{D}^2+d^2+\mathrm{D}d)$
$=\frac{1}{3}h(\mathrm{A}+a+\sqrt{\mathrm{A}\times a})$.

Prism

Volume $=$ area of base $\times$ height.

Wedge

Volume $=\frac{1}{6}$(length of edge $+2$ length of back)bh.

Prismoidal Formula

The prismoidal formula enables the volume of a prism, pyramid, or frustum of a pyramid to be found.

$$\text{Volume}=\frac{\text{end areas}+4\times\text{mid area}}{6}\times\text{height}.$$

Sphere

Surface $=d^2\pi=4\pi r^2$

Volume $=\dfrac{d^3\pi}{6}=\dfrac{4}{3}\pi r^3$

Segment of Sphere

$r=$ rad. of base.

Volume $=\dfrac{\pi}{6}h(3r^2+h^2)$.

$r=$ rad. of sphere.

Volume $=\dfrac{\pi}{3}h^2(3r-h)$.

Cube

Volume $=$ length $\times$ breadth $\times$ height.

Spherical Zone

Volume $=\dfrac{\pi}{2}h(\frac{1}{3}h^2+\mathrm{R}^2+r^2)$.

Surface area of convex part of segment or zone of sphere
$=\pi d$(of sphere)h
$=2\pi rh$.

Mid-spherical zone :

Volume $=(r+\dfrac{2}{3}h^2)\dfrac{\pi}{4}$.

Spheroid

Volume=revolving axis^2 × fixed axis × $\frac{\pi}{6}$.

Cylinder

Area$=2\pi r^2+\pi dh$.
Volume$=\pi r^2 h$.

Torus or Solid Ring

Volume$=2\pi^2 \mathrm{R}r^2$
$=19{\cdot}74\mathrm{R}r^2$
$=2{\cdot}463\mathrm{D}d^2$.

UNITS

Velocity, Acceleration, Force, Energy, and Power

Derived Units.—Derived units are units for the measurement of quantities which are based upon the fundamental units.

Absolute Units.—These are derived units referred directly to the fundamental units of mass, length, and time.

Fundamental Units.—The fundamental units are the units of length, mass, and time.

Arbitrary Units.—These are arbitrary combined units, not related to the fundamental units ; thus the weight of 1 pound is an arbitary unit because it depends upon the locality of the earth and upon the size and density of the earth.

Velocity.—This is the ratio of displacement in a specified direction to the time occupied.

Speed is the same as velocity, but with no reference to direction.

At any instant the linear velocity of a particle is the rate at which its position is changing. The unit of velocity is a rate of change of position of 1 ft. per second, or 1 centimetre per second. Velocity is expressed by length divided by time, and the dimensions of a velocity are therefore length divided by time. This is expressed by the dimensional equation $V=LT^{-1}$.

Acceleration is the rate at which velocity is changing at any instant. The dimensions of acceleration are LT^{-2} or $\frac{L}{T^2}$ and are expressed in feet or centimetres per second per second. On the C.G.S. system, that is in centimetres and seconds, acceleration of gravity is 981 centimetres per second per second, and the acceleration of gravity in feet per second per second is $\frac{981}{30\cdot5}=32\cdot2$.

Momentum is the product of mass and velocity at any instant. Its dimensions are therefore $\frac{ML}{T}=MLT^{-1}$.

Force is measured by the rate at which the momentum of a body is changing at the particular instant. The dimensions of force are $\frac{ML}{T^2}$ or MLT^{-2}.

Newton's Laws of Motion.—(1) Every material body persists in a state of rest or uniform motion straight line unless acted upon by an external force.

(2) The time-rate of change of momentum is proportional to the force applied and takes place in the direction in which the force acts.

(3) To every action there is always an equal and opposite reaction.

The Dyne.—The unit of force on the centimetre-gramme-second system is the dyne. It is the force which after acting on a mass of 1 gramme for 1 second gives it a velocity of 1 centimetre per second.

The Poundal.—The unit of force on the foot-pound system is the poundal. A force of 1 dyne, acting on a mass of 1 gm. imparts to it an acceleration of 1 cm. per sec. per sec. The poundal is an absolute unit of force on the f.p.s. system, but the weight of 1 pound is a gravitation and arbitrary unit and varies with the locality.

Moment of Inertia.—The moment of inertia round any axis is defined as the sum of each element of mass of the body, each multiplied by the square of its perpendicular distance from the axis, thus the angular velocity is $\frac{d\theta}{dt}$. The angular acceleration is $\frac{d^2\theta}{dt^2}$. The angular momentum is $\frac{d\theta}{Kdt}$ and the torque or angular force or impressed couple is $\frac{d^2\theta}{Kdt^2}$.

Energy or Work.—When a body is displaced against the action of a force tending to move it in the opposite direction, work is said to be done or energy expended on it. In the case of a mass m raised to height h against the uniform acceleration of gravity g, work equal to mgh units is done.

The Foot-Pound.—The unit of work in mechanical engineering is the foot-pound, and it is the work done in lifting 1 pound 1 foot against gravity. This, however, is dependent on the acceleration of gravity at the locality.

The Foot-Poundal.—The absolute unit of work in the foot-pound system is the foot-poundal, which is nearly equal to $\frac{1}{32 \cdot 2}$ foot-pounds.

The Centimetre-Dyne or Erg.—The unit of work on the C.G.S. system is a centimetre-dyne, which is called the erg. Ten million ergs are called 1 joule. The dimensions of work are those of force multiplied by length or ML^2T^{-2}.

The Horse-Power Hour.—This is a common engineering unit of work. One horse-power equals a rate of doing work of 33,000 foot-pounds per minute, or 550 foot-pounds per second. A horse-power hour is the amount of work done when 1 horse-power continues for one hour. It is therefore equal to 1,980,000 foot-pounds. It is therefore a gravitation unit and depends on locality. As 1 foot-pound is approximately equal to 32·2 foot-poundals, 1 horse-power hour is equal to 63·75 times 10^6 foot-poundals, which equals 1,980,000 times 32·2.

The dimensions of power are those of work divided by time, which equals ML^2T^{-3}. The change ratio from f.p.s. to c.g.s. is therefore 421,390.

Power.—Power is the rate of doing work. The c.g.s. unit of power is therefore 1 erg per second. The f.p.s. unit of power is 1 foot-poundal per second.

The Kip is the kilo pound, or 1000 lb.

ELECTRICAL UNITS

The volt is the practical unit of electromotive force, or difference of potential or electrical pressure.

The ohm is the practical unit of resistance, which varies directly as the length and inversely as the area of section of a conductor.

The ampere is the practical unit of strength of current, or velocity. It is often described as the measure of current density.

The coulomb is the practical unit of quantity, and represents the amount of electricity given by 1 ampere in 1 second. (The term " coulomb " is becoming obsolete.)

The farad is the unit of capacity of an electrical condenser ; one-millionth of this, or the micro-farad, is taken as the practical unit. A condenser of 1 farad capacity would be raised to the potential of 1 volt by the charge of 1 coulomb of electricity.

The watt is the practical unit of work, and is the amount of work required to force 1 ampere through 1 ohm during 1 second.

The joule is the mechanical equivalent of heat, or the work done in generating one heat unit or calorie, and is equal to 1 watt per second. The watt-hour joule$=10^7$, or 10,000,000 ergs. The Board of Trade Unit equals 3,600,000 joules, and is known as the Kilowatt-hour. Named after Joule, the English physicist. Joule's Law states that the heat produced by a current I passing through a resistance R for time t is proportional to I^2Rt ; 1 calorie$=4{\cdot}2\times10^7$ ergs, or 4·2 joules.

The henry is the unit of inductance or the coefficient of self-induction.

The volt may be understood as a measure of pressure, the ampere of quantity, the watt of power ; thus a current of 10 amperes at 10 volts$=$100 watts.

The Board of Trade Commercial Unit$=$1 kelvin$=$3,600,000 joules$=$1000 watt-hours$=$1 kilowatt per hour$=$1·34 H.P. for 1 hour. It equals a current of 1000 amperes at an E.M.F. of 1 volt flowing for 1 hour. For private lighting it may be taken as 10 A 100 V, and for public lighting 5 A 200 V.

Measure of Electrical Work

A=strength of current in amperes.
V=electromotive force in volts.
O=resistance in ohms.
C=quantity of electricity in coulombs.
t=time in seconds.
$H.P.$=actual horse-power.
W=units of work or watts (1 unit=10 million ergs =absolute C.G.S. measurement).

$$A=\frac{V}{O}=\frac{C}{t},\ C=At.$$

$$H.P.\frac{AV}{746}=\frac{A^2O}{746}=\frac{W}{746}.$$

$$W=AV=A^2O.$$

1 watt $\frac{1}{746}$ of a H.P.=1 volt ampere=10^7 ergs per second.
1 kilowatt=1000 watts=10^{10} ergs per second.
1 kilowatt-hour=1·34 H.P. acting for 1 hour=say, $2\frac{3}{4}$ million foot-lb.

Electrical Equations

Amperes × volts=watts.
Joules ÷ seconds=watts.
Coulombs per second=amperes.
Watts ÷ 746=effective H.P.
Coulombs ÷ volts=farads.
0·7373 foot-lb. per second=1 joule.
Volts × coulombs=joules.
Watts × 44·236=foot-lb. per minute.
Kilowatts × 1·34=H.P.

An erg is the work done by 1 dyne acting through 1 centimetre. A dyne is $\frac{1}{981}$ of a gramme. A gramme is the weight of a cubic centimetre of pure water at 4 deg. C.=0·00220462 lb. 1 erg=0·00000007373229 foot-lb., and 1 million ergs=0·07, etc., foot-lb.; 10^7 ergs=10 million ergs=0·7, etc., foot-lb. =1 joule.

Ohm's Law

A law which gives the relations existing in any circuit between current, voltage, and resistance. The formula is :

current=voltage÷resistance, which is set down in mathematical form thus : $I=\frac{E}{R}$ (I being the electrical symbol for current, E the symbol for voltage, and R the symbol for resistance). From this equation it is obvious that the voltage can be found by multiplying the current by the resistance ($E=I\times R$); and the resistance is given by dividing the voltage by the current $\left(R=\frac{E}{I}\right)$. In all the above equations the three terms must be in the units of the respective measurements, namely, I in amperes, E in volts, and R in ohms. For an example, take a circuit consisting of a battery of 6 volts, across which is joined a resistance of 3 ohms, and this results in a current of 2 amps.

Current$=\frac{6}{3}=2$ amps.

Resistance$=\frac{6}{2}=3$ ohms.

Voltage$=2\times 3=6$ volts.

Ohms' Law for A.C.—Circuits having inductance : $I=\frac{E}{2\pi fL}$.

For A.C. circuits having capacity only, the formula is : $I=2\pi fcE$, or $E=\frac{I}{2\pi fC}$. Where f=frequency in cycles per second, C=capacity, V=voltage, and L=inductance in Henrys. Expressed another way $I=\frac{E}{Z}$, where Z=impedance of circuit.

NATURAL SINES, COSINES, AND TANGENTS

Degrees.	Sine.	Cosine.	Tangent.	Degrees.	Sine.	Cosine.	Tangent.
30′	·0087	·9999	·0087	23° 0′	·3907	·9205	0·4245
1° 0′	·0174	·9998	·0174	30′	·3987	·9170	0·4348
30′	·0262	·9997	·0262	24° 0′	·4067	·9136	0·4452
2° 0′	·0349	·9994	·0349	30′	·4147	·9099	0·4557
30′	·0436	·9991	·0437	25° 0′	·4226	·9063	0·4663
3° 0′	·0523	·9986	·0524	30′	·4305	·9025	0·4770
30′	·0610	·9981	·0612	26° 0′	·4384	·8988	0·4877
4° 0′	·0698	·9976	·0699	30′	·4462	·8949	0·4986
30′	·0785	·9969	·0787	27° 0′	·4540	·8910	0·5095
5° 0′	·0872	·9962	·0875	30′	·4617	·8870	0·5206
30′	·0959	·9954	·0963	28° 0′	·4695	·8829	0·5317
6° 0′	·1045	·9945	·1051	30′	·4772	·8788	0·5430
30′	·1132	·9936	·1139	29° 0′	·4818	·8746	0·5543
7° 0′	·1219	·9926	·1228	30′	·4924	·8704	0·5658
30′	·1305	·9914	·1317	30° 0′	·5000	·8660	0·5774
8° 0′	·1392	·9903	·1405	30′	·5075	·8616	0·5891
30′	·1478	·9890	·1495	31° 0′	·5150	·8572	0·6009
9° 0′	·1564	·9877	·1584	30′	·5225	·8526	0·6128
30′	·1650	·9863	·1673	32° 0′	·5299	·8480	0·6249
10° 0′	·1737	·9848	·1763	30′	·5373	·8434	0·6371
30′	·1822	·9833	·1853	33° 0′	·5446	·8387	0·6494
11° 0′	·1908	·9816	·1944	30′	·5519	·8339	0·6619
30′	·1994	·9800	·2035	34° 0′	·5592	·8290	0·6745
12° 0′	·2079	·9782	·2126	30′	·5664	·8241	0·6873
30′	·2164	·9763	·2217	35° 0′	·5736	·8192	0·7002
13° 0′	·2249	·9744	·2309	30′	·5807	·8142	0·7133
30′	·2334	·9724	·2401	36° 0′	·5878	·8090	0·7265
14° 0′	·2419	·9703	·2493	30′	·5948	·8039	0·7400
30′	·2504	·9682	·2586	37° 0′	·6018	·7986	0·7536
15° 0′	·2588	·9659	·2679	30′	·6088	·7934	0·7673
30′	·2672	·9636	·2773	38° 0′	·6157	·7880	0·7813
16° 0′	·2756	·9613	·2867	30′	·6225	·7826	0·7954
30′	·2840	·9588	·2962	39° 0′	·6293	·7772	0·8098
17° 0′	·2924	·9563	·3057	30′	·6361	·7716	0·8243
30′	·3007	·9537	·3153	40° 0′	·6428	·7660	0·8391
18° 0′	·3090	·9510	·3249	30′	·6494	·7604	0·8541
30′	·3173	·9483	·3346	41° 0′	·6561	·7547	0·8693
19° 0′	·3256	·9455	·3443	30′	·6626	·7490	0·8847
30′	·3338	·9426	·3541	42° 0′	·6691	·7431	0·9004
20° 0′	·3420	·9397	·3640	30′	·6756	·7373	0·9163
30′	·3502	·9367	·3739	43° 0′	·6820	·7314	0·9325
21° 0′	·3584	·9336	·3839	30′	·6884	·7254	0·9489
30′	·3665	·9304	·3939	44° 0′	·6947	·7193	0·9657
22° 0′	·3746	·9272	·4040	30′	·7009	·7132	0·9827
30′	·3827	·9239	·4142	45° 0′	·7071	·7071	1·0000

NATURAL SINES, COSINES, AND TANGENTS—contd.

Degrees.	Sine.	Cosine.	Tangent.	Degrees.	Sine.	Cosine.	Tangent.
45° 30′	·7133	·7009	1·0176	68° 0′	0·9272	·3746	2·4751
46° 0′	·7193	·6947	1·0355	30′	0·9304	·3665	2·5387
30′	·7254	·6884	1·0538	69° 0′	0·9336	·3584	2·6051
47° 0′	·7314	·6820	1·0724	30′	0·9367	·3502	2·6746
30′	·7373	·6756	1·0913	70° 0′	0·9397	·3420	2·7475
48° 0′	·7431	·6691	1·1106	30′	0·9426	·3338	2·8239
30′	·7490	·6626	1·1303	71° 0′	0·9455	·3256	2·9042
49° 0′	·7547	·6561	1·1504	30′	0·9483	·3173	2·9887
30′	·7604	·6495	1·1709	72° 0′	0·9510	·3090	3·0777
50° 0′	·7660	·6428	1·1918	30′	0·9537	·3007	3·1716
30′	·7716	·6361	1·2131	73° 0′	0·9563	·2924	3·2907
51° 0′	·7771	·6293	1·2349	30′	0·9588	·2840	3·3759
30′	·7826	·6225	1·2572	74° 0′	0·9613	·2756	3·4874
52° 0′	·7880	·6157	1·2799	30′	0·9636	·2672	3·6059
30′	·7934	·6088	1·3032	75° 0′	0·9659	·2588	3·7321
53° 0′	·7986	·6018	1·3270	30′	0·9682	·2504	3·8667
30′	·8039	·5948	1·3514	76° 0′	0·9703	·2419	4·0108
54° 0′	·8090	·5878	1·3764	30′	0·9724	·2335	4·1653
30′	·8141	·5807	1·4020	77° 0′	0·9744	·2249	4·3315
55° 0′	·8192	·5736	1·4282	30′	0·9763	·2164	4·5107
30′	·8241	·5664	1·4550	78° 0′	0·9782	·2079	4·7046
56° 0′	·8290	·5592	1·4826	30′	0·9800	·1994	4·9152
30′	·8339	·5519	1·5108	79° 0′	0·9816	·1908	5·1446
57° 0′	·8387	·5446	1·5399	30′	0·9833	·1822	5·3955
30′	·8434	·5373	1·5697	80° 0′	0·9848	·1737	5·6713
58° 0′	·8480	·5299	1·6003	30′	0·9863	·1650	5·9758
30′	·8526	·5225	1·6319	81° 0′	0·9877	·1564	6·3138
59° 0′	·8572	·5150	1·6643	30′	0·9890	·1478	6·6912
30′	·8616	·5075	1·6977	82° 0′	0·9903	·1392	7·1154
60° 0′	·8660	·5000	1·7321	30′	0·9914	·1305	7·5958
30′	·8704	·4924	1·7675	83° 0′	0·9926	·1219	8·1444
61° 0′	·8746	·4848	1·8041	30′	0·9936	·1132	8·7769
30′	·8788	·4772	1·8418	84° 0′	0·9945	·1045	9·5144
62° 0′	·8830	·4695	1·8807	30′	0·9954	·0959	10·3854
30′	·8870	·4618	1·9210	85° 0′	0·9962	·0872	11·4301
63° 0′	·8910	·4540	1·9626	30′	0·9969	·0785	12·7062
30′	·8949	·4462	2·0057	86° 0′	0·9976	·0698	14·3007
64° 0′	·8988	·4384	2·0503	30′	0·9981	·0610	16·3499
30′	·9026	·4305	2·0965	87° 0′	0·9986	·0523	19·0811
65° 0′	·9063	·4226	2·1445	30′	0·9990	·0436	22·9038
30′	·9100	·4147	2·1943	88° 0′	0·9994	·0349	28·6363
66° 0′	·9135	·4067	2·2460	30′	0·9997	·0262	38·1885
30′	·9171	·3988	2·2998	89° 0′	0·9998	·0174	57·2900
67° 0′	·9205	·3907	2·3559	30′	0·9999	·0087	114·5887
30′	·9239	·3827	2·4142	90° 0′	1·0000	·0000	∞

Logarithms

	0	1	2	3	4	5	6	7	8	9	1	2	3	4	5	6	7	8	9
10	0000	0043	0086	0128	0170	0212	0253	0294	0334	0374	4	8	12	17	21	25	29	33	37
11	0414	0453	0492	0531	0569	0607	0645	0682	0719	0755	4	8	11	15	19	23	26	30	34
12	0792	0828	0864	0899	0934	0969	1004	1038	1072	1106	3	7	10	14	17	21	24	28	31
13	1139	1173	1206	1239	1271	1303	1335	1367	1399	1430	3	6	10	13	16	19	23	26	29
14	1461	1492	1523	1553	1584	1614	1644	1673	1703	1732	3	6	9	12	15	18	21	24	27
15	1761	1790	1818	1847	1875	1903	1931	1959	1987	2014	3	6	8	11	14	17	20	22	25
16	2041	2068	2095	2122	2148	2175	2201	2227	2253	2279	3	5	8	11	13	16	18	21	24
17	2304	2330	2355	2380	2405	2430	2455	2480	2504	2529	2	5	7	10	12	15	17	20	22
18	2553	2577	2601	2625	2648	2672	2695	2718	2742	2765	2	5	7	9	12	14	16	19	21
19	2788	2810	2833	2856	2878	2900	2923	2945	2967	2989	2	4	7	9	11	13	16	18	20
20	3010	3032	3054	3075	3096	3118	3139	3160	3181	3201	2	4	6	8	11	13	15	17	19
21	3222	3243	3263	3284	3304	3324	3345	3365	3385	3404	2	4	6	8	10	12	14	16	18
22	3424	3444	3464	3483	3502	3522	3541	3560	3579	3598	2	4	6	8	10	12	14	15	17
23	3617	3636	3655	3674	3692	3711	3729	3747	3766	3784	2	4	6	7	9	11	13	15	17
24	3802	3820	3838	3856	3874	3892	3909	3927	3945	3962	2	4	5	7	9	11	12	14	16
25	3979	3997	4014	4031	4048	4065	4082	4099	4116	4133	2	3	5	7	9	10	12	14	15
26	4150	4166	4183	4200	4216	4232	4249	4265	4281	4298	2	3	5	7	8	10	11	13	15
27	4314	4330	4346	4362	4378	4393	4409	4425	4440	4456	2	3	5	6	8	9	11	13	14
28	4472	4487	4502	4518	4533	4548	4564	4579	4594	4609	2	3	5	6	8	9	11	12	14
29	4624	4639	4654	4669	4683	4698	4713	4728	4742	4757	1	3	4	6	7	9	10	12	13

30	4771	4786	4800	4814	4829	4843	4857	4871	4886	4900	1	3	4	6	7	9	10	11	13
31	4914	4928	4942	4955	4969	4983	4997	5011	5024	5038	1	3	4	6	7	8	10	11	12
32	5051	5065	5079	5092	5105	5119	5132	5145	5159	5172	1	3	4	5	7	8	9	11	12
33	5185	5198	5211	5224	5237	5250	5263	5276	5289	5302	1	3	4	5	6	8	9	10	12
34	5315	5328	5340	5353	5366	5378	5391	5403	5416	5428	1	3	4	5	6	8	9	10	11
35	5441	5453	5465	5478	5490	5502	5514	5527	5539	5551	1	2	4	5	6	7	9	10	11
36	5563	5575	5587	5599	5611	5623	5635	5647	5658	5670	1	2	4	5	6	7	8	10	11
37	5682	5694	5705	5717	5729	5740	5752	5763	5775	5786	1	2	3	5	6	7	8	9	10
38	5798	5809	5821	5832	5843	5855	5866	5877	5888	5899	1	2	3	5	6	7	8	9	10
39	5911	5922	5933	5944	5955	5966	5977	5988	5999	6010	1	2	3	4	5	7	8	9	10
40	6021	6031	6042	6053	6064	6075	6085	6096	6107	6117	1	2	3	4	5	6	8	9	10
41	6128	6138	6149	6160	6170	6180	6191	6201	6212	6222	1	2	3	4	5	6	7	8	9
42	6232	6243	6253	6263	6274	6284	6294	6304	6314	6325	1	2	3	4	5	6	7	8	9
43	6335	6345	6355	6365	6375	6385	6395	6405	6415	6425	1	2	3	4	5	6	7	8	9
44	6435	6444	6454	6464	6474	6484	6493	6503	6513	6522	1	2	3	4	5	6	7	8	9
45	6532	6542	6551	6561	6571	6580	6590	6599	6609	6618	1	2	3	4	5	6	7	8	9
46	6628	6637	6646	6656	6665	6675	6684	6693	6702	6712	1	2	3	4	5	6	7	7	8
47	6721	6730	6739	6749	6758	6767	6776	6785	6794	6803	1	2	3	4	5	5	6	7	8
48	6812	6821	6830	6839	6848	6857	6866	6875	6884	6893	1	2	3	4	4	5	6	7	8
49	6902	6911	6920	6928	6937	6946	6955	6964	6972	6981	1	2	3	4	4	5	6	7	8
50	6990	6998	7007	7016	7024	7033	7042	7050	7059	7067	1	2	3	3	4	5	6	7	8
51	7076	7084	7093	7101	7110	7118	7126	7135	7143	7152	1	2	3	3	4	5	6	7	8
52	7160	7168	7177	7185	7193	7202	7210	7218	7226	7235	1	2	2	3	4	5	6	7	7
53	7243	7251	7259	7267	7275	7284	7292	7300	7308	7316	1	2	2	3	4	5	6	6	7
54	7324	7332	7340	7348	7356	7364	7372	7380	7388	7396	1	2	2	3	4	5	6	6	7

Logarithms—continued

	0	1	2	3	4	5	6	7	8	9	1	2	3	4	5	6	7	8	9
55	7404	7412	7419	7427	7435	7443	7451	7459	7466	7474	1	2	2	3	4	5	5	6	7
56	7482	7490	7497	7505	7513	7520	7528	7536	7543	7551	1	2	2	3	4	5	5	6	7
57	7559	7566	7574	7582	7589	7597	7604	7612	7619	7627	1	2	2	3	4	5	5	6	7
58	7634	7642	7649	7657	7664	7672	7679	7686	7694	7701	1	1	2	3	4	4	5	6	7
59	7709	7716	7723	7731	7738	7745	7752	7760	7767	7774	1	1	2	3	4	4	5	6	7
60	7782	7789	7796	7803	7810	7818	7825	7832	7839	7846	1	1	2	3	4	4	5	6	6
61	7853	7860	7868	7875	7882	7889	7896	7903	7910	7917	1	1	2	3	4	4	5	6	6
62	7924	7931	7938	7945	7952	7959	7966	7973	7980	7987	1	1	2	3	3	4	5	6	6
63	7993	8000	8007	8014	8021	8028	8035	8041	8048	8055	1	1	2	3	3	4	5	5	6
64	8062	8069	8075	8082	8089	8096	8102	8109	8116	8122	1	1	2	3	3	4	5	5	6
65	8129	8136	8142	8149	8156	8162	8169	8176	8182	8189	1	1	2	3	3	4	5	5	6
66	8195	8202	8209	8215	8222	8228	8235	8241	8248	8254	1	1	2	3	3	4	5	5	6
67	8261	8267	8274	8280	8287	8293	8299	8306	8312	8319	1	1	2	3	3	4	5	5	6
68	8325	8331	8338	8344	8351	8357	8363	8370	8376	8382	1	1	2	3	3	4	4	5	6
69	8388	8395	8401	8407	8414	8420	8426	8432	8439	8445	1	1	2	2	3	4	4	5	6
70	8451	8457	8463	8470	8476	8482	8488	8494	8500	8506	1	1	2	2	3	4	4	5	6
71	8513	8519	8525	8531	8537	8543	8549	8555	8561	8567	1	1	2	2	3	4	4	5	5
72	8573	8579	8585	8591	8597	8603	8609	8615	8621	8627	1	1	2	2	3	4	4	5	5
73	8633	8639	8645	8651	8657	8663	8669	8675	8681	8686	1	1	2	2	3	4	4	5	5
74	8692	8698	8704	8710	8716	8722	8727	8733	8739	8745	1	1	2	2	3	4	4	5	5

75	8751	8756	8762	8768	8774	8779	8785	8791	8797	8802	1	1	2	2	3	3	4	5	5
76	8808	8814	8820	8825	8831	8837	8842	8848	8854	8859	1	1	2	2	3	3	4	5	5
77	8865	8871	8876	8882	8887	8893	8899	8904	8910	8915	1	1	2	2	3	3	4	4	5
78	8921	8927	8932	8938	8943	8949	8954	8960	8965	8971	1	1	2	2	3	3	4	4	5
79	8976	8982	8987	8993	8998	9004	9009	9015	9020	9025	1	1	2	2	3	3	4	4	5
80	9031	9036	9042	9047	9053	9058	9063	9069	9074	9079	1	1	2	2	3	3	4	4	5
81	9085	9090	9096	9101	9106	9112	9117	9122	9128	9133	1	1	2	2	3	3	4	4	5
82	9138	9143	9149	9154	9159	9165	9170	9175	9180	9186	1	1	2	2	3	3	4	4	5
83	9191	9196	9201	9206	9212	9217	9222	9227	9232	9238	1	1	2	2	3	3	4	4	5
84	9243	9248	9253	9258	9263	9269	9274	9279	9284	9289	1	1	2	2	3	3	4	4	5
85	9294	9299	9304	9309	9315	9320	9325	9330	9335	9340	1	1	2	2	3	3	4	4	5
86	9345	9350	9355	9360	9365	9370	9375	9380	9385	9390	1	1	2	2	3	3	4	4	5
87	9395	9400	9405	9410	9415	9420	9425	9430	9435	9440	0	1	1	2	2	3	3	4	4
88	9445	9450	9455	9460	9465	9469	9474	9479	9484	9489	0	1	1	2	2	3	3	4	4
89	9494	9499	9504	9509	9513	9518	9523	9528	9533	9538	0	1	1	2	2	3	3	4	4
90	9542	9547	9552	9557	9562	9566	9571	9576	9581	9586	0	1	1	2	2	3	3	4	4
91	9590	9595	9600	9605	9609	9614	9619	9624	9628	9633	0	1	1	2	2	3	3	4	4
92	9638	9643	9647	9652	9657	9661	9666	9671	9675	9680	0	1	1	2	2	3	3	4	4
93	9685	9689	9694	9699	9703	9708	9713	9717	9722	9727	0	1	1	2	2	3	3	4	4
94	9731	9736	9741	9745	9750	9754	9759	9763	9768	9773	0	1	1	2	2	3	3	4	4
95	9777	9782	9786	9791	9795	9800	9805	9809	9814	9818	0	1	1	2	2	3	3	4	4
96	9823	9827	9832	9836	9841	9845	9850	9854	9859	9863	0	1	1	2	2	3	3	4	4
97	9868	9872	9877	9881	9886	9890	9894	9899	9903	9908	0	1	1	2	2	3	3	4	4
98	9912	9917	9921	9926	9930	9934	9939	9943	9948	9952	0	1	1	2	2	3	3	4	4
99	9956	9961	9965	9969	9974	9978	9983	9987	9991	9996	0	1	1	2	2	3	3	3	4

Antilogarithms

	0	1	2	3	4	5	6	7	8	9	1	2	3	4	5	6	7	8	9
·00	1000	1002	1005	1007	1009	1012	1014	1016	1019	1021	0	0	1	1	1	1	2	2	2
·01	1023	1026	1028	1030	1033	1035	1038	1040	1042	1045	0	0	1	1	1	1	2	2	2
·02	1047	1050	1052	1054	1057	1059	1062	1064	1067	1069	0	0	1	1	1	1	2	2	2
·03	1072	1074	1076	1079	1081	1084	1086	1089	1091	1094	0	0	1	1	1	1	2	2	2
·04	1096	1099	1102	1104	1107	1109	1112	1114	1117	1119	0	1	1	1	1	2	2	2	2
·05	1122	1125	1127	1130	1132	1135	1138	1140	1143	1146	0	1	1	1	1	2	2	2	2
·06	1148	1151	1153	1156	1159	1161	1164	1167	1169	1172	0	1	1	1	1	2	2	2	2
·07	1175	1178	1180	1183	1186	1189	1191	1194	1197	1199	0	1	1	1	1	2	2	2	2
·08	1202	1205	1208	1211	1213	1216	1219	1222	1225	1227	0	1	1	1	1	2	2	2	3
·09	1230	1233	1236	1239	1242	1245	1247	1250	1253	1256	0	1	1	1	1	2	2	2	3
·10	1259	1262	1265	1268	1271	1274	1276	1279	1282	1285	0	1	1	1	1	2	2	2	3
·11	1288	1291	1294	1297	1300	1303	1306	1309	1312	1315	0	1	1	1	2	2	2	2	3
·12	1318	1321	1324	1327	1330	1334	1337	1340	1343	1346	0	1	1	1	2	2	2	2	3
·13	1349	1352	1355	1358	1361	1365	1368	1371	1374	1377	0	1	1	1	2	2	2	3	3
·14	1380	1384	1387	1390	1393	1396	1400	1403	1406	1409	0	1	1	1	2	2	2	3	3
·15	1413	1416	1419	1422	1426	1429	1432	1435	1439	1442	0	1	1	1	2	2	2	3	3
·16	1445	1449	1452	1455	1459	1462	1466	1469	1472	1476	0	1	1	1	2	2	2	3	3
·17	1479	1483	1486	1489	1493	1496	1500	1503	1507	1510	0	1	1	1	2	2	2	3	3
·18	1514	1517	1521	1524	1528	1531	1535	1538	1542	1545	0	1	1	1	2	2	2	3	3
·19	1549	1552	1556	1560	1563	1567	1570	1574	1578	1581	0	1	1	1	2	2	3	3	3
·20	1585	1589	1592	1596	1600	1603	1607	1611	1614	1618	0	1	1	1	2	2	3	3	3
·21	1622	1626	1629	1633	1637	1641	1644	1648	1652	1656	0	1	1	2	2	2	3	3	3

·22	1660	1663	1667	1671	1675	1679	1683	1687	1690	1694	0	1	1	2	2	2	3	3	3
·23	1698	1702	1706	1710	1714	1718	1722	1726	1730	1734	0	1	1	2	2	2	3	3	4
·24	1738	1742	1746	1750	1754	1758	1762	1766	1770	1774	0	1	1	2	2	2	3	3	4
·25	1778	1782	1786	1791	1795	1799	1803	1807	1811	1816	0	1	1	2	2	2	3	3	4
·26	1820	1824	1828	1832	1837	1841	1845	1849	1854	1858	0	1	1	2	2	3	3	3	4
·27	1862	1866	1871	1875	1879	1884	1888	1892	1897	1901	0	1	1	2	2	3	3	3	4
·28	1905	1910	1914	1919	1923	1928	1932	1936	1941	1945	0	1	1	2	2	3	3	4	4
·29	1950	1954	1959	1963	1968	1972	1977	1982	1986	1991	0	1	1	2	2	3	3	4	4
·30	1995	2000	2004	2009	2014	2018	2023	2028	2032	2037	0	1	1	2	2	3	3	4	4
·31	2042	2046	2051	2056	2061	2065	2070	2075	2080	2084	0	1	1	2	2	3	3	4	4
·32	2089	2094	2099	2104	2109	2113	2118	2123	2128	2133	0	1	1	2	2	3	3	4	4
·33	2138	2143	2148	2153	2158	2163	2168	2173	2178	2183	0	1	1	2	2	3	3	4	4
·34	2188	2193	2198	2203	2208	2213	2218	2223	2228	2234	1	1	2	2	3	3	4	4	5
·35	2239	2244	2249	2254	2259	2265	2270	2275	2280	2286	1	1	2	2	3	3	4	4	5
·36	2291	2296	2301	2307	2312	2317	2323	2328	2333	2339	1	1	2	2	3	3	4	4	5
·37	2344	2350	2355	2360	2366	2371	2377	2382	2388	2393	1	1	2	2	3	3	4	4	5
·38	2399	2404	2410	2415	2421	2427	2432	2438	2443	2449	1	1	2	2	3	3	4	4	5
·39	2455	2460	2466	2472	2477	2483	2489	2495	2500	2506	1	1	2	2	3	3	4	5	5
·40	2512	2518	2523	2529	2535	2541	2547	2553	2559	2564	1	1	2	2	3	4	4	5	5
·41	2570	2576	2582	2588	2594	2600	2606	2612	2618	2624	1	1	2	2	3	4	4	5	5
·42	2630	2636	2642	2649	2655	2661	2667	2673	2679	2685	1	1	2	2	3	4	4	5	6
·43	2692	2698	2704	2710	2716	2723	2729	2735	2742	2748	1	1	2	3	3	4	4	5	6
·44	2754	2761	2767	2773	2780	2786	2793	2799	2805	2812	1	1	2	3	3	4	4	5	6
·45	2818	2825	2831	2838	2844	2851	2858	2864	2871	2877	1	1	2	3	3	4	5	5	6
·46	2884	2891	2897	2904	2911	2917	2924	2931	2938	2944	1	1	2	3	3	4	5	5	6
·47	2951	2958	2965	2972	2979	2985	2992	2999	3006	3013	1	1	2	3	3	4	5	5	6
·48	3020	3027	3034	3041	3048	3055	3062	3069	3076	3083	1	1	2	3	4	4	5	6	6
·49	3090	3097	3105	3112	3119	3126	3133	3141	3148	3155	1	1	2	3	4	4	5	6	6

Antilogarithms—continued

	0	1	2	3	4	5	6	7	8	9	1	2	3	4	5	6	7	8	9
·50	3162	3170	3177	3184	3192	3199	3206	3214	3221	3228	1	1	2	3	4	4	5	6	7
·51	3236	3243	3251	3258	3266	3273	3281	3289	3296	3304	1	2	2	3	4	5	5	6	7
·52	3311	3319	3327	3334	3342	3350	3357	3365	3373	3381	1	2	2	3	4	5	5	6	7
·53	3388	3396	3404	3412	3420	3428	3436	3443	3451	3459	1	2	2	3	4	5	6	6	7
·54	3467	3475	3483	3491	3499	3508	3516	3524	3532	3540	1	2	2	3	4	5	6	6	7
·55	3548	3556	3565	3573	3581	3589	3597	3606	3614	3622	1	2	2	3	4	5	6	7	7
·56	3631	3639	3648	3656	3664	3673	3681	3690	3698	3707	1	2	3	3	4	5	6	7	8
·57	3715	3724	3733	3741	3750	3758	3767	3776	3784	3793	1	2	3	3	4	5	6	7	8
·58	3802	3811	3819	3828	3837	3846	3855	3864	3873	3882	1	2	3	4	4	5	6	7	8
·59	3890	3899	3908	3917	3926	3936	3945	3954	3963	3972	1	2	3	4	5	5	6	7	8
·60	3981	3990	3999	4009	4018	4027	4036	4046	4055	4064	1	2	3	4	5	6	6	7	8
·61	4074	4083	4093	4102	4111	4121	4130	4140	4150	4159	1	2	3	4	5	6	7	8	9
·62	4169	4178	4188	4198	4207	4217	4227	4236	4246	4256	1	2	3	4	5	6	7	8	9
·63	4266	4276	4285	4295	4305	4315	4325	4335	4345	4355	1	2	3	4	5	6	7	8	9
·64	4365	4375	4385	4395	4406	4416	4426	4436	4446	4457	1	2	3	4	5	6	7	8	9
·65	4467	4477	4487	4498	4508	4519	4529	4539	4550	4560	1	2	3	4	5	6	7	8	9
·66	4571	4581	4592	4603	4613	4624	4634	4645	4656	4667	1	2	3	4	5	6	7	9	10
·67	4677	4688	4699	4710	4721	4732	4742	4753	4764	4775	1	2	3	4	5	7	8	9	10
·68	4786	4797	4808	4819	4831	4842	4853	4864	4875	4887	1	2	3	4	6	7	8	9	10
·69	4898	4909	4920	4932	4943	4955	4966	4977	4989	5000	1	2	3	5	6	7	8	9	10
·70	5012	5023	5035	5047	5058	5070	5082	5093	5105	5117	1	2	4	5	6	7	8	9	11
·71	5129	5140	5152	5164	5176	5188	5200	5212	5224	5236	1	2	4	5	6	7	8	10	11

·72	5248	5260	5272	5284	5297	5309	5321	5333	5346	5358	1	2	4	5	6	7	9	10	11
·73	5370	5383	5395	5408	5420	5433	5445	5458	5470	5483	1	3	4	5	6	8	9	10	11
·74	5495	5508	5521	5534	5546	5559	5572	5585	5598	5610	1	3	4	5	6	8	9	10	12
·75	5623	5636	5649	5662	5675	5689	5702	5715	5728	5741	1	3	4	5	7	8	9	10	12
·76	5754	5768	5781	5794	5808	5821	5834	5848	5861	5875	1	3	4	5	7	8	9	11	12
·77	5888	5902	5916	5929	5943	5957	5970	5984	5998	6012	1	3	4	5	7	8	10	11	12
·78	6026	6039	6053	6067	6081	6095	6109	6124	6138	6152	1	3	4	6	7	8	10	11	13
·79	6166	6180	6194	6209	6223	6237	6252	6266	6281	6295	1	3	4	6	7	9	10	11	13
·80	6310	6324	6339	6353	6368	6383	6397	6412	6427	6442	1	3	4	6	7	9	10	12	13
·81	6457	6471	6486	6501	6516	6531	6546	6561	6577	6592	2	3	5	6	8	9	11	12	14
·82	6607	6622	6637	6653	6668	6683	6699	6714	6730	6745	2	3	5	6	8	9	11	12	14
·83	6761	6776	6792	6808	6823	6839	6855	6871	6887	6902	2	3	5	6	8	9	11	13	14
·84	6918	6934	6950	6966	6982	6998	7015	7031	7047	7063	2	3	5	6	8	10	11	13	15
·85	7079	7096	7112	7129	7145	7161	7178	7194	7211	7228	2	3	5	7	8	10	12	13	15
·86	7244	7261	7278	7295	7311	7328	7345	7362	7379	7396	2	3	5	7	8	10	12	13	15
·87	7413	7430	7447	7464	7482	7499	7516	7534	7551	7568	2	4	5	7	9	10	12	14	16
·88	7586	7603	7621	7638	7656	7674	7691	7709	7727	7745	2	4	5	7	9	11	12	14	16
·89	7762	7780	7798	7816	7834	7852	7870	7889	7907	7925	2	4	5	7	9	11	13	14	16
·90	7943	7962	7980	7998	8017	8035	8054	8072	8091	8110	2	4	6	7	9	11	13	15	17
·91	8128	8147	8166	8185	8204	8222	8241	8260	8279	8299	2	4	6	8	9	11	13	15	17
·92	8318	8337	8356	8375	8395	8414	8433	8453	8472	8492	2	4	6	8	10	12	14	15	17
·93	8511	8531	8551	8570	8590	8610	8630	8650	8670	8690	2	4	6	8	10	12	14	16	18
·94	8710	8730	8750	8770	8790	8810	8831	8851	8872	8892	2	4	6	8	10	12	14	16	18
·95	8913	8933	8954	8974	8995	9016	9036	9057	9078	9099	2	4	6	8	10	12	15	17	19
·96	9120	9141	9162	9183	9204	9226	9247	9268	9290	9311	2	4	6	8	11	13	15	17	19
·97	9333	9354	9376	9397	9419	9441	9462	9484	9506	9528	2	4	7	9	11	13	15	17	20
·98	9550	9572	9594	9616	9638	9661	9683	9705	9727	9750	2	4	7	9	11	13	16	18	20
·99	9772	9795	9817	9840	9863	9886	9908	9931	9954	9977	2	5	7	9	11	14	16	18	20

INDEX

B

C

E

F

G

H

I

T

U

V

W

Y

Z

A CATALOG OF SELECTED

DOVER BOOKS

IN ALL FIELDS OF INTEREST

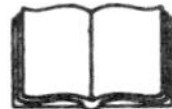

A CATALOG OF SELECTED DOVER BOOKS IN ALL FIELDS OF INTEREST

CONCERNING THE SPIRITUAL IN ART, Wassily Kandinsky. Pioneering work by father of abstract art. Thoughts on color theory, nature of art. Analysis of earlier masters. 12 illustrations. 80pp. of text. 5⅜ x 8½. 23411-8

ANIMALS: 1,419 Copyright-Free Illustrations of Mammals, Birds, Fish, Insects, etc., Jim Harter (ed.). Clear wood engravings present, in extremely lifelike poses, over 1,000 species of animals. One of the most extensive pictorial sourcebooks of its kind. Captions. Index. 284pp. 9 x 12. 23766-4

CELTIC ART: The Methods of Construction, George Bain. Simple geometric techniques for making Celtic interlacements, spirals, Kells-type initials, animals, humans, etc. Over 500 illustrations. 160pp. 9 x 12. (Available in U.S. only.) 22923-8

AN ATLAS OF ANATOMY FOR ARTISTS, Fritz Schider. Most thorough reference work on art anatomy in the world. Hundreds of illustrations, including selections from works by Vesalius, Leonardo, Goya, Ingres, Michelangelo, others. 593 illustrations. 192pp. 7⅛ x 10¼. 20241-0

CELTIC HAND STROKE-BY-STROKE (Irish Half-Uncial from "The Book of Kells"): An Arthur Baker Calligraphy Manual, Arthur Baker. Complete guide to creating each letter of the alphabet in distinctive Celtic manner. Covers hand position, strokes, pens, inks, paper, more. Illustrated. 48pp. 8¼ x 11. 24336-2

EASY ORIGAMI, John Montroll. Charming collection of 32 projects (hat, cup, pelican, piano, swan, many more) specially designed for the novice origami hobbyist. Clearly illustrated easy-to-follow instructions insure that even beginning papercrafters will achieve successful results. 48pp. 8¼ x 11. 27298-2

THE COMPLETE BOOK OF BIRDHOUSE CONSTRUCTION FOR WOODWORKERS, Scott D. Campbell. Detailed instructions, illustrations, tables. Also data on bird habitat and instinct patterns. Bibliography. 3 tables. 63 illustrations in 15 figures. 48pp. 5¼ x 8½. 24407-5

BLOOMINGDALE'S ILLUSTRATED 1886 CATALOG: Fashions, Dry Goods and Housewares, Bloomingdale Brothers. Famed merchants' extremely rare catalog depicting about 1,700 products: clothing, housewares, firearms, dry goods, jewelry, more. Invaluable for dating, identifying vintage items. Also, copyright-free graphics for artists, designers. Co-published with Henry Ford Museum & Greenfield Village. 160pp. 8¼ x 11. 25780-0

HISTORIC COSTUME IN PICTURES, Braun & Schneider. Over 1,450 costumed figures in clearly detailed engravings–from dawn of civilization to end of 19th century. Captions. Many folk costumes. 256pp. 8⅝ x 11¾. 23150-X

STICKLEY CRAFTSMAN FURNITURE CATALOGS, Gustav Stickley and L. & J. G. Stickley. Beautiful, functional furniture in two authentic catalogs from 1910. 594 illustrations, including 277 photos, show settles, rockers, armchairs, reclining chairs, bookcases, desks, tables. 183pp. 6½ x 9¼. 23838-5

AMERICAN LOCOMOTIVES IN HISTORIC PHOTOGRAPHS: 1858 to 1949, Ron Ziel (ed.). A rare collection of 126 meticulously detailed official photographs, called "builder portraits," of American locomotives that majestically chronicle the rise of steam locomotive power in America. Introduction. Detailed captions. xi+ 129pp. 9 x 12. 27393-8

AMERICA'S LIGHTHOUSES: An Illustrated History, Francis Ross Holland, Jr. Delightfully written, profusely illustrated fact-filled survey of over 200 American lighthouses since 1716. History, anecdotes, technological advances, more. 240pp. 8 x 10¾. 25576-X

TOWARDS A NEW ARCHITECTURE, Le Corbusier. Pioneering manifesto by founder of "International School." Technical and aesthetic theories, views of industry, economics, relation of form to function, "mass-production split" and much more. Profusely illustrated. 320pp. 6⅛ x 9¼. (Available in U.S. only.) 25023-7

HOW THE OTHER HALF LIVES, Jacob Riis. Famous journalistic record, exposing poverty and degradation of New York slums around 1900, by major social reformer. 100 striking and influential photographs. 233pp. 10 x 7⅞. 22012-5

FRUIT KEY AND TWIG KEY TO TREES AND SHRUBS, William M. Harlow. One of the handiest and most widely used identification aids. Fruit key covers 120 deciduous and evergreen species; twig key 160 deciduous species. Easily used. Over 300 photographs. 126pp. 5⅜ x 8½. 20511-8

COMMON BIRD SONGS, Dr. Donald J. Borror. Songs of 60 most common U.S. birds: robins, sparrows, cardinals, bluejays, finches, more–arranged in order of increasing complexity. Up to 9 variations of songs of each species.
Cassette and manual 99911-4

ORCHIDS AS HOUSE PLANTS, Rebecca Tyson Northen. Grow cattleyas and many other kinds of orchids–in a window, in a case, or under artificial light. 63 illustrations. 148pp. 5⅜ x 8½. 23261-1

MONSTER MAZES, Dave Phillips. Masterful mazes at four levels of difficulty. Avoid deadly perils and evil creatures to find magical treasures. Solutions for all 32 exciting illustrated puzzles. 48pp. 8¼ x 11. 26005-4

MOZART'S DON GIOVANNI (DOVER OPERA LIBRETTO SERIES), Wolfgang Amadeus Mozart. Introduced and translated by Ellen H. Bleiler. Standard Italian libretto, with complete English translation. Convenient and thoroughly portable–an ideal companion for reading along with a recording or the performance itself. Introduction. List of characters. Plot summary. 121pp. 5¼ x 8½. 24944-1

TECHNICAL MANUAL AND DICTIONARY OF CLASSICAL BALLET, Gail Grant. Defines, explains, comments on steps, movements, poses and concepts. 15-page pictorial section. Basic book for student, viewer. 127pp. 5⅜ x 8½. 21843-0

THE CLARINET AND CLARINET PLAYING, David Pino. Lively, comprehensive work features suggestions about technique, musicianship, and musical interpretation, as well as guidelines for teaching, making your own reeds, and preparing for public performance. Includes an intriguing look at clarinet history. "A godsend," *The Clarinet,* Journal of the International Clarinet Society. Appendixes. 7 illus. 320pp. 5⅜ x 8½. 40270-3

HOLLYWOOD GLAMOR PORTRAITS, John Kobal (ed.). 145 photos from 1926-49. Harlow, Gable, Bogart, Bacall; 94 stars in all. Full background on photographers, technical aspects. 160pp. 8⅜ x 11¼. 23352-9

THE ANNOTATED CASEY AT THE BAT: A Collection of Ballads about the Mighty Casey/Third, Revised Edition, Martin Gardner (ed.). Amusing sequels and parodies of one of America's best-loved poems: Casey's Revenge, Why Casey Whiffed, Casey's Sister at the Bat, others. 256pp. 5⅜ x 8½. 28598-7

THE RAVEN AND OTHER FAVORITE POEMS, Edgar Allan Poe. Over 40 of the author's most memorable poems: "The Bells," "Ulalume," "Israfel," "To Helen," "The Conqueror Worm," "Eldorado," "Annabel Lee," many more. Alphabetic lists of titles and first lines. 64pp. 5$\frac{3}{16}$ x 8¼. 26685-0

PERSONAL MEMOIRS OF U. S. GRANT, Ulysses Simpson Grant. Intelligent, deeply moving firsthand account of Civil War campaigns, considered by many the finest military memoirs ever written. Includes letters, historic photographs, maps and more. 528pp. 6⅛ x 9¼. 28587-1

ANCIENT EGYPTIAN MATERIALS AND INDUSTRIES, A. Lucas and J. Harris. Fascinating, comprehensive, thoroughly documented text describes this ancient civilization's vast resources and the processes that incorporated them in daily life, including the use of animal products, building materials, cosmetics, perfumes and incense, fibers, glazed ware, glass and its manufacture, materials used in the mummification process, and much more. 544pp. 6⅛ x 9¼. (Available in U.S. only.) 40446-3

RUSSIAN STORIES/RUSSKIE RASSKAZY: A Dual-Language Book, edited by Gleb Struve. Twelve tales by such masters as Chekhov, Tolstoy, Dostoevsky, Pushkin, others. Excellent word-for-word English translations on facing pages, plus teaching and study aids, Russian/English vocabulary, biographical/critical introductions, more. 416pp. 5⅜ x 8½. 26244-8

PHILADELPHIA THEN AND NOW: 60 Sites Photographed in the Past and Present, Kenneth Finkel and Susan Oyama. Rare photographs of City Hall, Logan Square, Independence Hall, Betsy Ross House, other landmarks juxtaposed with contemporary views. Captures changing face of historic city. Introduction. Captions. 128pp. 8¼ x 11. 25790-8

AIA ARCHITECTURAL GUIDE TO NASSAU AND SUFFOLK COUNTIES, LONG ISLAND, The American Institute of Architects, Long Island Chapter, and the Society for the Preservation of Long Island Antiquities. Comprehensive, well-researched and generously illustrated volume brings to life over three centuries of Long Island's great architectural heritage. More than 240 photographs with authoritative, extensively detailed captions. 176pp. 8¼ x 11. 26946-9

NORTH AMERICAN INDIAN LIFE: Customs and Traditions of 23 Tribes, Elsie Clews Parsons (ed.). 27 fictionalized essays by noted anthropologists examine religion, customs, government, additional facets of life among the Winnebago, Crow, Zuni, Eskimo, other tribes. 480pp. 6⅛ x 9¼. 27377-6

FRANK LLOYD WRIGHT'S DANA HOUSE, Donald Hoffmann. Pictorial essay of residential masterpiece with over 160 interior and exterior photos, plans, elevations, sketches and studies. 128pp. 9¼ x 10¾. 29120-0

THE MALE AND FEMALE FIGURE IN MOTION: 60 Classic Photographic Sequences, Eadweard Muybridge. 60 true-action photographs of men and women walking, running, climbing, bending, turning, etc., reproduced from rare 19th-century masterpiece. vi + 121pp. 9 x 12. 24745-7

1001 QUESTIONS ANSWERED ABOUT THE SEASHORE, N. J. Berrill and Jacquelyn Berrill. Queries answered about dolphins, sea snails, sponges, starfish, fishes, shore birds, many others. Covers appearance, breeding, growth, feeding, much more. 305pp. 5¼ x 8¼. 23366-9

ATTRACTING BIRDS TO YOUR YARD, William J. Weber. Easy-to-follow guide offers advice on how to attract the greatest diversity of birds: birdhouses, feeders, water and waterers, much more. 96pp. $5\frac{3}{16}$ x 8¼. 28927-3

MEDICINAL AND OTHER USES OF NORTH AMERICAN PLANTS: A Historical Survey with Special Reference to the Eastern Indian Tribes, Charlotte Erichsen-Brown. Chronological historical citations document 500 years of usage of plants, trees, shrubs native to eastern Canada, northeastern U.S. Also complete identifying information. 343 illustrations. 544pp. 6½ x 9¼. 25951-X

STORYBOOK MAZES, Dave Phillips. 23 stories and mazes on two-page spreads: Wizard of Oz, Treasure Island, Robin Hood, etc. Solutions. 64pp. 8¼ x 11. 23628-5

AMERICAN NEGRO SONGS: 230 Folk Songs and Spirituals, Religious and Secular, John W. Work. This authoritative study traces the African influences of songs sung and played by black Americans at work, in church, and as entertainment. The author discusses the lyric significance of such songs as "Swing Low, Sweet Chariot," "John Henry," and others and offers the words and music for 230 songs. Bibliography. Index of Song Titles. 272pp. 6½ x 9¼. 40271-1

MOVIE-STAR PORTRAITS OF THE FORTIES, John Kobal (ed.). 163 glamor, studio photos of 106 stars of the 1940s: Rita Hayworth, Ava Gardner, Marlon Brando, Clark Gable, many more. 176pp. 8⅜ x 11¼. 23546-7

BENCHLEY LOST AND FOUND, Robert Benchley. Finest humor from early 30s, about pet peeves, child psychologists, post office and others. Mostly unavailable elsewhere. 73 illustrations by Peter Arno and others. 183pp. 5⅜ x 8½. 22410-4

YEKL and THE IMPORTED BRIDEGROOM AND OTHER STORIES OF YIDDISH NEW YORK, Abraham Cahan. Film Hester Street based on *Yekl* (1896). Novel, other stories among first about Jewish immigrants on N.Y.'s East Side. 240pp. 5⅜ x 8½. 22427-9

SELECTED POEMS, Walt Whitman. Generous sampling from *Leaves of Grass.* Twenty-four poems include "I Hear America Singing," "Song of the Open Road," "I Sing the Body Electric," "When Lilacs Last in the Dooryard Bloom'd," "O Captain! My Captain!"–all reprinted from an authoritative edition. Lists of titles and first lines. 128pp. $5\frac{3}{16}$ x 8¼. 26878-0

MY BONDAGE AND MY FREEDOM, Frederick Douglass. Born a slave, Douglass became outspoken force in antislavery movement. The best of Douglass' autobiographies. Graphic description of slave life. 464pp. 5⅜ x 8½. 22457-0

FOLLOWING THE EQUATOR: A Journey Around the World, Mark Twain. Fascinating humorous account of 1897 voyage to Hawaii, Australia, India, New Zealand, etc. Ironic, bemused reports on peoples, customs, climate, flora and fauna, politics, much more. 197 illustrations. 720pp. 5⅜ x 8½. 26113-1

THE PEOPLE CALLED SHAKERS, Edward D. Andrews. Definitive study of Shakers: origins, beliefs, practices, dances, social organization, furniture and crafts, etc. 33 illustrations. 351pp. 5⅜ x 8½. 21081-2

THE MYTHS OF GREECE AND ROME, H. A. Guerber. A classic of mythology, generously illustrated, long prized for its simple, graphic, accurate retelling of the principal myths of Greece and Rome, and for its commentary on their origins and significance. With 64 illustrations by Michelangelo, Raphael, Titian, Rubens, Canova, Bernini and others. 480pp. 5⅜ x 8½. 27584-1

PSYCHOLOGY OF MUSIC, Carl E. Seashore. Classic work discusses music as a medium from psychological viewpoint. Clear treatment of physical acoustics, auditory apparatus, sound perception, development of musical skills, nature of musical feeling, host of other topics. 88 figures. 408pp. 5⅜ x 8½. 21851-1

THE PHILOSOPHY OF HISTORY, Georg W. Hegel. Great classic of Western thought develops concept that history is not chance but rational process, the evolution of freedom. 457pp. 5⅜ x 8½. 20112-0

THE BOOK OF TEA, Kakuzo Okakura. Minor classic of the Orient: entertaining, charming explanation, interpretation of traditional Japanese culture in terms of tea ceremony. 94pp. 5⅜ x 8½. 20070-1

LIFE IN ANCIENT EGYPT, Adolf Erman. Fullest, most thorough, detailed older account with much not in more recent books, domestic life, religion, magic, medicine, commerce, much more. Many illustrations reproduce tomb paintings, carvings, hieroglyphs, etc. 597pp. 5⅜ x 8½. 22632-8

SUNDIALS, Their Theory and Construction, Albert Waugh. Far and away the best, most thorough coverage of ideas, mathematics concerned, types, construction, adjusting anywhere. Simple, nontechnical treatment allows even children to build several of these dials. Over 100 illustrations. 230pp. 5⅜ x 8½. 22947-5

THEORETICAL HYDRODYNAMICS, L. M. Milne-Thomson. Classic exposition of the mathematical theory of fluid motion, applicable to both hydrodynamics and aerodynamics. Over 600 exercises. 768pp. 6⅛ x 9¼. 68970-0

SONGS OF EXPERIENCE: Facsimile Reproduction with 26 Plates in Full Color, William Blake. 26 full-color plates from a rare 1826 edition. Includes "The Tyger," "London," "Holy Thursday," and other poems. Printed text of poems. 48pp. 5¼ x 7. 24636-1

OLD-TIME VIGNETTES IN FULL COLOR, Carol Belanger Grafton (ed.). Over 390 charming, often sentimental illustrations, selected from archives of Victorian graphics–pretty women posing, children playing, food, flowers, kittens and puppies, smiling cherubs, birds and butterflies, much more. All copyright-free. 48pp. 9¼ x 12¼. 27269-9

PERSPECTIVE FOR ARTISTS, Rex Vicat Cole. Depth, perspective of sky and sea, shadows, much more, not usually covered. 391 diagrams, 81 reproductions of drawings and paintings. 279pp. 5⅜ x 8½. 22487-2

DRAWING THE LIVING FIGURE, Joseph Sheppard. Innovative approach to artistic anatomy focuses on specifics of surface anatomy, rather than muscles and bones. Over 170 drawings of live models in front, back and side views, and in widely varying poses. Accompanying diagrams. 177 illustrations. Introduction. Index. 144pp. 8⅜ x11¼. 26723-7

GOTHIC AND OLD ENGLISH ALPHABETS: 100 Complete Fonts, Dan X. Solo. Add power, elegance to posters, signs, other graphics with 100 stunning copyright-free alphabets: Blackstone, Dolbey, Germania, 97 more–including many lower-case, numerals, punctuation marks. 104pp. 8⅛ x 11. 24695-7

HOW TO DO BEADWORK, Mary White. Fundamental book on craft from simple projects to five-bead chains and woven works. 106 illustrations. 142pp. 5⅜ x 8. 20697-1

THE BOOK OF WOOD CARVING, Charles Marshall Sayers. Finest book for beginners discusses fundamentals and offers 34 designs. "Absolutely first rate . . . well thought out and well executed."–E. J. Tangerman. 118pp. 7¾ x 10⅝. 23654-4

ILLUSTRATED CATALOG OF CIVIL WAR MILITARY GOODS: Union Army Weapons, Insignia, Uniform Accessories, and Other Equipment, Schuyler, Hartley, and Graham. Rare, profusely illustrated 1846 catalog includes Union Army uniform and dress regulations, arms and ammunition, coats, insignia, flags, swords, rifles, etc. 226 illustrations. 160pp. 9 x 12. 24939-5

WOMEN'S FASHIONS OF THE EARLY 1900s: An Unabridged Republication of "New York Fashions, 1909," National Cloak & Suit Co. Rare catalog of mail-order fashions documents women's and children's clothing styles shortly after the turn of the century. Captions offer full descriptions, prices. Invaluable resource for fashion, costume historians. Approximately 725 illustrations. 128pp. 8⅜ x 11¼. 27276-1

THE 1912 AND 1915 GUSTAV STICKLEY FURNITURE CATALOGS, Gustav Stickley. With over 200 detailed illustrations and descriptions, these two catalogs are essential reading and reference materials and identification guides for Stickley furniture. Captions cite materials, dimensions and prices. 112pp. 6½ x 9¼. 26676-1

EARLY AMERICAN LOCOMOTIVES, John H. White, Jr. Finest locomotive engravings from early 19th century: historical (1804–74), main-line (after 1870), special, foreign, etc. 147 plates. 142pp. 11⅜ x 8¼. 22772-3

THE TALL SHIPS OF TODAY IN PHOTOGRAPHS, Frank O. Braynard. Lavishly illustrated tribute to nearly 100 majestic contemporary sailing vessels: Amerigo Vespucci, Clearwater, Constitution, Eagle, Mayflower, Sea Cloud, Victory, many more. Authoritative captions provide statistics, background on each ship. 190 black-and-white photographs and illustrations. Introduction. 128pp. 8⅞ x 11¾. 27163-3

LITTLE BOOK OF EARLY AMERICAN CRAFTS AND TRADES, Peter Stockham (ed.). 1807 children's book explains crafts and trades: baker, hatter, cooper, potter, and many others. 23 copperplate illustrations. 140pp. 4⅝ x 6. 23336-7

VICTORIAN FASHIONS AND COSTUMES FROM HARPER'S BAZAR, 1867–1898, Stella Blum (ed.). Day costumes, evening wear, sports clothes, shoes, hats, other accessories in over 1,000 detailed engravings. 320pp. 9⅜ x 12¼. 22990-4

GUSTAV STICKLEY, THE CRAFTSMAN, Mary Ann Smith. Superb study surveys broad scope of Stickley's achievement, especially in architecture. Design philosophy, rise and fall of the Craftsman empire, descriptions and floor plans for many Craftsman houses, more. 86 black-and-white halftones. 31 line illustrations. Introduction 208pp. 6½ x 9¼. 27210-9

THE LONG ISLAND RAIL ROAD IN EARLY PHOTOGRAPHS, Ron Ziel. Over 220 rare photos, informative text document origin (1844) and development of rail service on Long Island. Vintage views of early trains, locomotives, stations, passengers, crews, much more. Captions. 8⅞ x 11¾. 26301-0

VOYAGE OF THE LIBERDADE, Joshua Slocum. Great 19th-century mariner's thrilling, first-hand account of the wreck of his ship off South America, the 35-foot boat he built from the wreckage, and its remarkable voyage home. 128pp. 5⅜ x 8½. 40022-0

TEN BOOKS ON ARCHITECTURE, Vitruvius. The most important book ever written on architecture. Early Roman aesthetics, technology, classical orders, site selection, all other aspects. Morgan translation. 331pp. 5⅜ x 8½. 20645-9

THE HUMAN FIGURE IN MOTION, Eadweard Muybridge. More than 4,500 stopped-action photos, in action series, showing undraped men, women, children jumping, lying down, throwing, sitting, wrestling, carrying, etc. 390pp. 7⅞ x 10⅝. 20204-6 Clothbd.

TREES OF THE EASTERN AND CENTRAL UNITED STATES AND CANADA, William M. Harlow. Best one-volume guide to 140 trees. Full descriptions, woodlore, range, etc. Over 600 illustrations. Handy size. 288pp. 4½ x 6⅜. 20395-6

SONGS OF WESTERN BIRDS, Dr. Donald J. Borror. Complete song and call repertoire of 60 western species, including flycatchers, juncoes, cactus wrens, many more–includes fully illustrated booklet. Cassette and manual 99913-0

GROWING AND USING HERBS AND SPICES, Milo Miloradovich. Versatile handbook provides all the information needed for cultivation and use of all the herbs and spices available in North America. 4 illustrations. Index. Glossary. 236pp. 5⅜ x 8½. 25058-X

BIG BOOK OF MAZES AND LABYRINTHS, Walter Shepherd. 50 mazes and labyrinths in all–classical, solid, ripple, and more–in one great volume. Perfect inexpensive puzzler for clever youngsters. Full solutions. 112pp. 8⅛ x 11. 22951-3

PIANO TUNING, J. Cree Fischer. Clearest, best book for beginner, amateur. Simple repairs, raising dropped notes, tuning by easy method of flattened fifths. No previous skills needed. 4 illustrations. 201pp. 5⅜ x 8½. 23267-0

HINTS TO SINGERS, Lillian Nordica. Selecting the right teacher, developing confidence, overcoming stage fright, and many other important skills receive thoughtful discussion in this indispensible guide, written by a world-famous diva of four decades' experience. 96pp. 5⅜ x 8½. 40094-8

THE COMPLETE NONSENSE OF EDWARD LEAR, Edward Lear. All nonsense limericks, zany alphabets, Owl and Pussycat, songs, nonsense botany, etc., illustrated by Lear. Total of 320pp. 5⅜ x 8½. (Available in U.S. only.) 20167-8

VICTORIAN PARLOUR POETRY: An Annotated Anthology, Michael R. Turner. 117 gems by Longfellow, Tennyson, Browning, many lesser-known poets. "The Village Blacksmith," "Curfew Must Not Ring Tonight," "Only a Baby Small," dozens more, often difficult to find elsewhere. Index of poets, titles, first lines. xxiii + 325pp. 5⅝ x 8¼. 27044-0

DUBLINERS, James Joyce. Fifteen stories offer vivid, tightly focused observations of the lives of Dublin's poorer classes. At least one, "The Dead," is considered a masterpiece. Reprinted complete and unabridged from standard edition. 160pp. 5³⁄₁₆ x 8¼. 26870-5

GREAT WEIRD TALES: 14 Stories by Lovecraft, Blackwood, Machen and Others, S. T. Joshi (ed.). 14 spellbinding tales, including "The Sin Eater," by Fiona McLeod, "The Eye Above the Mantel," by Frank Belknap Long, as well as renowned works by R. H. Barlow, Lord Dunsany, Arthur Machen, W. C. Morrow and eight other masters of the genre. 256pp. 5⅜ x 8½. (Available in U.S. only.) 40436-6

THE BOOK OF THE SACRED MAGIC OF ABRAMELIN THE MAGE, translated by S. MacGregor Mathers. Medieval manuscript of ceremonial magic. Basic document in Aleister Crowley, Golden Dawn groups. 268pp. 5⅜ x 8½. 23211-5

NEW RUSSIAN-ENGLISH AND ENGLISH-RUSSIAN DICTIONARY, M. A. O'Brien. This is a remarkably handy Russian dictionary, containing a surprising amount of information, including over 70,000 entries. 366pp. 4½ x 6⅛. 20208-9

HISTORIC HOMES OF THE AMERICAN PRESIDENTS, Second, Revised Edition, Irvin Haas. A traveler's guide to American Presidential homes, most open to the public, depicting and describing homes occupied by every American President from George Washington to George Bush. With visiting hours, admission charges, travel routes. 175 photographs. Index. 160pp. 8¼ x 11. 26751-2

NEW YORK IN THE FORTIES, Andreas Feininger. 162 brilliant photographs by the well-known photographer, formerly with *Life* magazine. Commuters, shoppers, Times Square at night, much else from city at its peak. Captions by John von Hartz. 181pp. 9¼ x 10¾. 23585-8

INDIAN SIGN LANGUAGE, William Tomkins. Over 525 signs developed by Sioux and other tribes. Written instructions and diagrams. Also 290 pictographs. 111pp. 6⅛ x 9¼. 22029-X

ANATOMY: A Complete Guide for Artists, Joseph Sheppard. A master of figure drawing shows artists how to render human anatomy convincingly. Over 460 illustrations. 224pp. 8⅜ x 11¼. 27279-6

MEDIEVAL CALLIGRAPHY: Its History and Technique, Marc Drogin. Spirited history, comprehensive instruction manual covers 13 styles (ca. 4th century through 15th). Excellent photographs; directions for duplicating medieval techniques with modern tools. 224pp. 8⅜ x 11¼. 26142-5

DRIED FLOWERS: How to Prepare Them, Sarah Whitlock and Martha Rankin. Complete instructions on how to use silica gel, meal and borax, perlite aggregate, sand and borax, glycerine and water to create attractive permanent flower arrangements. 12 illustrations. 32pp. 5⅜ x 8½. 21802-3

EASY-TO-MAKE BIRD FEEDERS FOR WOODWORKERS, Scott D. Campbell. Detailed, simple-to-use guide for designing, constructing, caring for and using feeders. Text, illustrations for 12 classic and contemporary designs. 96pp. 5⅜ x 8½. 25847-5

SCOTTISH WONDER TALES FROM MYTH AND LEGEND, Donald A. Mackenzie. 16 lively tales tell of giants rumbling down mountainsides, of a magic wand that turns stone pillars into warriors, of gods and goddesses, evil hags, powerful forces and more. 240pp. 5⅜ x 8½. 29677-6

THE HISTORY OF UNDERCLOTHES, C. Willett Cunnington and Phyllis Cunnington. Fascinating, well-documented survey covering six centuries of English undergarments, enhanced with over 100 illustrations: 12th-century laced-up bodice, footed long drawers (1795), 19th-century bustles, 19th-century corsets for men, Victorian "bust improvers," much more. 272pp. 5⅝ x 8¼. 27124-2

ARTS AND CRAFTS FURNITURE: The Complete Brooks Catalog of 1912, Brooks Manufacturing Co. Photos and detailed descriptions of more than 150 now very collectible furniture designs from the Arts and Crafts movement depict davenports, settees, buffets, desks, tables, chairs, bedsteads, dressers and more, all built of solid, quarter-sawed oak. Invaluable for students and enthusiasts of antiques, Americana and the decorative arts. 80pp. 6½ x 9¼. 27471-3

WILBUR AND ORVILLE: A Biography of the Wright Brothers, Fred Howard. Definitive, crisply written study tells the full story of the brothers' lives and work. A vividly written biography, unparalleled in scope and color, that also captures the spirit of an extraordinary era. 560pp. 6⅛ x 9¼. 40297-5

THE ARTS OF THE SAILOR: Knotting, Splicing and Ropework, Hervey Garrett Smith. Indispensable shipboard reference covers tools, basic knots and useful hitches; handsewing and canvas work, more. Over 100 illustrations. Delightful reading for sea lovers. 256pp. 5⅜ x 8½. 26440-8

FRANK LLOYD WRIGHT'S FALLINGWATER: The House and Its History, Second, Revised Edition, Donald Hoffmann. A total revision–both in text and illustrations–of the standard document on Fallingwater, the boldest, most personal architectural statement of Wright's mature years, updated with valuable new material from the recently opened Frank Lloyd Wright Archives. "Fascinating"–*The New York Times.* 116 illustrations. 128pp. 9¼ x 10¾. 27430-6

PHOTOGRAPHIC SKETCHBOOK OF THE CIVIL WAR, Alexander Gardner. 100 photos taken on field during the Civil War. Famous shots of Manassas Harper's Ferry, Lincoln, Richmond, slave pens, etc. 244pp. 10⅝ x 8¼. 22731-6

FIVE ACRES AND INDEPENDENCE, Maurice G. Kains. Great back-to-the-land classic explains basics of self-sufficient farming. The one book to get. 95 illustrations. 397pp. 5⅜ x 8½. 20974-1

SONGS OF EASTERN BIRDS, Dr. Donald J. Borror. Songs and calls of 60 species most common to eastern U.S.: warblers, woodpeckers, flycatchers, thrushes, larks, many more in high-quality recording. Cassette and manual 99912-2

A MODERN HERBAL, Margaret Grieve. Much the fullest, most exact, most useful compilation of herbal material. Gigantic alphabetical encyclopedia, from aconite to zedoary, gives botanical information, medical properties, folklore, economic uses, much else. Indispensable to serious reader. 161 illustrations. 888pp. 6½ x 9¼. 2-vol. set. (Available in U.S. only.) Vol. I: 22798-7
Vol. II: 22799-5

HIDDEN TREASURE MAZE BOOK, Dave Phillips. Solve 34 challenging mazes accompanied by heroic tales of adventure. Evil dragons, people-eating plants, bloodthirsty giants, many more dangerous adversaries lurk at every twist and turn. 34 mazes, stories, solutions. 48pp. 8¼ x 11. 24566-7

LETTERS OF W. A. MOZART, Wolfgang A. Mozart. Remarkable letters show bawdy wit, humor, imagination, musical insights, contemporary musical world; includes some letters from Leopold Mozart. 276pp. 5⅜ x 8½. 22859-2

BASIC PRINCIPLES OF CLASSICAL BALLET, Agrippina Vaganova. Great Russian theoretician, teacher explains methods for teaching classical ballet. 118 illustrations. 175pp. 5⅜ x 8½. 22036-2

THE JUMPING FROG, Mark Twain. Revenge edition. The original story of The Celebrated Jumping Frog of Calaveras County, a hapless French translation, and Twain's hilarious "retranslation" from the French. 12 illustrations. 66pp. 5⅜ x 8½. 22686-7

BEST REMEMBERED POEMS, Martin Gardner (ed.). The 126 poems in this superb collection of 19th- and 20th-century British and American verse range from Shelley's "To a Skylark" to the impassioned "Renascence" of Edna St. Vincent Millay and to Edward Lear's whimsical "The Owl and the Pussycat." 224pp. 5⅜ x 8½. 27165-X

COMPLETE SONNETS, William Shakespeare. Over 150 exquisite poems deal with love, friendship, the tyranny of time, beauty's evanescence, death and other themes in language of remarkable power, precision and beauty. Glossary of archaic terms. 80pp. 5 3/16 x 8¼. 26686-9

THE BATTLES THAT CHANGED HISTORY, Fletcher Pratt. Eminent historian profiles 16 crucial conflicts, ancient to modern, that changed the course of civilization. 352pp. 5⅜ x 8½. 41129-X

THE WIT AND HUMOR OF OSCAR WILDE, Alvin Redman (ed.). More than 1,000 ripostes, paradoxes, wisecracks: Work is the curse of the drinking classes; I can resist everything except temptation; etc. 258pp. 5⅜ x 8½. 20602-5

SHAKESPEARE LEXICON AND QUOTATION DICTIONARY, Alexander Schmidt. Full definitions, locations, shades of meaning in every word in plays and poems. More than 50,000 exact quotations. 1,485pp. 6½ x 9¼. 2-vol. set.
Vol. 1: 22726-X
Vol. 2: 22727-8

SELECTED POEMS, Emily Dickinson. Over 100 best-known, best-loved poems by one of America's foremost poets, reprinted from authoritative early editions. No comparable edition at this price. Index of first lines. 64pp. 5$\frac{3}{16}$ x 8¼. 26466-1

THE INSIDIOUS DR. FU-MANCHU, Sax Rohmer. The first of the popular mystery series introduces a pair of English detectives to their archnemesis, the diabolical Dr. Fu-Manchu. Flavorful atmosphere, fast-paced action, and colorful characters enliven this classic of the genre. 208pp. 5$\frac{3}{16}$ x 8¼. 29898-1

THE MALLEUS MALEFICARUM OF KRAMER AND SPRENGER, translated by Montague Summers. Full text of most important witchhunter's "bible," used by both Catholics and Protestants. 278pp. 6⅝ x 10. 22802-9

SPANISH STORIES/CUENTOS ESPAÑOLES: A Dual-Language Book, Angel Flores (ed.). Unique format offers 13 great stories in Spanish by Cervantes, Borges, others. Faithful English translations on facing pages. 352pp. 5⅜ x 8½. 25399-6

GARDEN CITY, LONG ISLAND, IN EARLY PHOTOGRAPHS, 1869–1919, Mildred H. Smith. Handsome treasury of 118 vintage pictures, accompanied by carefully researched captions, document the Garden City Hotel fire (1899), the Vanderbilt Cup Race (1908), the first airmail flight departing from the Nassau Boulevard Aerodrome (1911), and much more. 96pp. 8⅞ x 11¾. 40669-5

OLD QUEENS, N.Y., IN EARLY PHOTOGRAPHS, Vincent F. Seyfried and William Asadorian. Over 160 rare photographs of Maspeth, Jamaica, Jackson Heights, and other areas. Vintage views of DeWitt Clinton mansion, 1939 World's Fair and more. Captions. 192pp. 8⅞ x 11. 26358-4

CAPTURED BY THE INDIANS: 15 Firsthand Accounts, 1750-1870, Frederick Drimmer. Astounding true historical accounts of grisly torture, bloody conflicts, relentless pursuits, miraculous escapes and more, by people who lived to tell the tale. 384pp. 5⅜ x 8½. 24901-8

THE WORLD'S GREAT SPEECHES (Fourth Enlarged Edition), Lewis Copeland, Lawrence W. Lamm, and Stephen J. McKenna. Nearly 300 speeches provide public speakers with a wealth of updated quotes and inspiration–from Pericles' funeral oration and William Jennings Bryan's "Cross of Gold Speech" to Malcolm X's powerful words on the Black Revolution and Earl of Spenser's tribute to his sister, Diana, Princess of Wales. 944pp. 5⅜ x 8⅜. 40903-1

THE BOOK OF THE SWORD, Sir Richard F. Burton. Great Victorian scholar/adventurer's eloquent, erudite history of the "queen of weapons"–from prehistory to early Roman Empire. Evolution and development of early swords, variations (sabre, broadsword, cutlass, scimitar, etc.), much more. 336pp. 6⅛ x 9¼.
25434-8